Aidos Musabekov
Dana Ermakhan
Madina Maulenkulova

MOLECULAR BIOLOGY

Aidos Musabekov
Dana Ermakhan
Madina Maulenkulova

MOLECULAR BIOLOGY

BIOLOGY

ScienciaScripts

Imprint
Any brand names and product names mentioned in this book are subject to trademark, brand or patent protection and are trademarks or registered trademarks of their respective holders. The use of brand names, product names, common names, trade names, product descriptions etc. even without a particular marking in this work is in no way to be construed to mean that such names may be regarded as unrestricted in respect of trademark and brand protection legislation and could thus be used by anyone.

Cover image: www.ingimage.com

This book is a translation from the original published under ISBN 978-620-4-20933-3.

Publisher:
Sciencia Scripts
is a trademark of
Dodo Books Indian Ocean Ltd. and OmniScriptum S.R.L publishing group

120 High Road, East Finchley, London, N2 9ED, United Kingdom
Str. Armeneasca 28/1, office 1, Chisinau MD-2012, Republic of Moldova, Europe
Managing Directors: Ieva Konstantinova, Victoria Ursu
info@omniscriptum.com

Printed at: see last page
ISBN: 978-620-8-51760-1

Contents

CHAPTER 1

1. Introduction to Molecular Biology Molecular Biology. Definition

The molecular biology course will focus on the study of nucleic acids and protein biosynthesis. Molecular biology is a science which, due to the regularities of interaction of polymeric molecules, leads to the occurrence of certain biological effects. That is, it provides functioning of biological systems due to intermolecular interactions. Speaking about interactions, it is necessary to know how molecules themselves are organised. There are different types of bonds (of different nature) in them.

Connections. Types of interactions

In biological molecules in particular, hydrogen bonds play a major role. The energy of hydrogen bonds is small. Thermal motion at physiological temperatures is sufficient to break a hydrogen bond. But if molecules form more than one such bond, then breaking one bond will not cause the molecules to separate. And, the more such bonds are formed, the more stable the complex will be. A large number of such bonds allows molecules to assemble into fairly stable complexes, but does not deprive the ability to make various rearrangements. Another type of interaction is Van der Waals interactions, which arise due to the attraction between atoms or molecules due to the fact that they have positive and negative charged components. The distance of the van der Waals radius determines the most compact stacking in the complex. Interactions of this kind are possible both between polar and non-polar molecules. In particular, hydrophobic interactions arise on their basis, when some molecules or fragments are more favourable to attract to each other, which as a result lead to the fact that if we have a lot of such points of contact, then, as in the case of hydrogen bonds, there are quite stable complexes. Hydrophobic interactions allow the formation of membranes.

Biological polymers, their types and the functions they perform fulfil

Biological molecules are mainly polymers. This makes it possible not to form each bond in each molecule anew, but to

to form rather large molecules quite quickly and economically. Large molecules are built from small blocks. Which as a rule are either the same or similar to each other. If a polymer is formed from molecules that have the same monomers in their composition, homopolymers (-A-A-A-A-A-A-) are obtained. They most often fulfil structural functions. Regular heteropolymers (-A-B-A-B-A-A-) are also limited in properties. These include cellulose, chitin (homopolymers), bacterial cell walls (regular heteropolymers) -

murein, hyaluronic acid (the basis of connective tissue), which also fulfil structural functions. The main ones are irregular heteropolymers, in which units do not obey a specific sequence (A-A-B-A-B-A-B-B-B-A-A-) and branched heteropolymers. The latter are quite compact and are used as storage matter - those substances that cells put aside for storage (glycogen in animals and fungi and starch in plants). If there is a dimer initially, then 4 polymer variants can be created from it, if there is a trimer - then 8, in general case M^n variants, where M is the number of monomers and n is the chain length of the polymer. Such heteropolymers have a strictly defined sequence of amino acids, so the process of their formation is of interest. Indeed, an enzyme is needed to recognise the sequence and attach the necessary monomers to the sequence. Enzymes are also proteins, and they too need to be synthesised. This problem was once called the problem of protein biosynthesis. To solve this problem, Koltsov proposed a mechanism that consisted of a matrix mechanism of synthesis. He knew that there was some structure called a chromosome on which the genes that determine certain traits were located. He suggested that chromosomes are where the special molecules of each protein sit, these protein molecules are there in a kind of unfolded form, and the special enzyme system that synthesises a new protein binds to the end of such an unfolded molecule and against the first amino acid binds the same for the next protein molecule. Then it moves for the next amino acid binds the same one and so on, i.e. it makes an exact copy of the original base molecule, which in this case is a matrix for synthesis of a new molecule. But this principle has not found application. Earlier it was believed that everything is built of proteins, Engels said: "life is the existence of protein bodies".

Discovery of nucleic acids. Composition and structure

However, the existence of carbohydrates and lipids was also known. Another type of polymeric molecules should be recalled. In the early 19th century, Friedrich Miescher discovered nucleic acids. He was studying the composition of humus. He collected bandages of purulent wounds in the hospital, treated them with certain solutions, including a solution obtained from the rennet of calves, where there was an enzyme pepsin, which destroys proteins, and after such treatment and separation of solid material, a fraction of leukocytes was practically obtained. Unlike other cells, leucocytes have very little cytoplasm. Their task during their short lifetime is to find and neutralise something. Their rather small amount of cytoplasm is quickly saturated by bacteria or toxic substances eaten and they die, but because they are normal cells from normal cellular precursors, they retain a cell nucleus. When Misher destroyed the leukocytes and studied their composition, he obtained a polymeric fraction, unlike other substances known at that time. It turned out that this substance was very high in phosphorus. The same substance he isolated from the milk of male salmon, other researchers then obtained it from brewer's yeast, bran. Since these were constituents of nuclei, they were called nucleins, and later, because of the manifestation of acidic properties - nucleic acids. It is worth noting that living cells have a slightly alkaline pH, so acids are not there, but their salts are. A group of researchers showed that it is a polymer consisting of complex monomers - nucleotides. The central element in a nucleotide is a five-carbon sugar: ribose or deosyribose, arbitrarily forming a beta-furanose form in solution (Fig. 1.1).

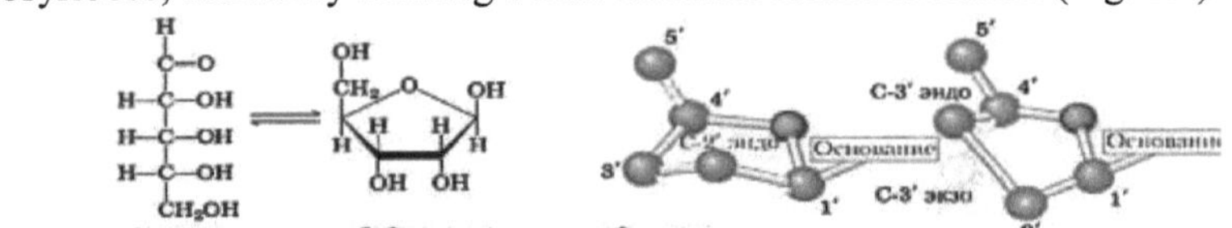

Fig. 1.1 furanose form of sugar

Two other components are attached to this major component. The first is a phosphate, which in free nucleotides is always attached to the fifth 7 to the last carbon due to the formation of an ester bond. Upon decomposition, there may be an attachment to the third position. If there is a hydroxyl in the second position of furanose, it is a ribose; if there is no hydroxyl, it is a deoxyribose. A purine or pyrimidine base takes the place of the hydroxyl in the first position (Fig. 1.2).

Figure 1.2. Structure of a nucleotide

Pyrimidines include cytosine, thymine, uracil, and purines include adenine and guanine. The formation of nucleotide mono-, di-, and triphosphate is possible. Most nucleotides are in the triphosphate form (Fig. 1.3, 1.4).

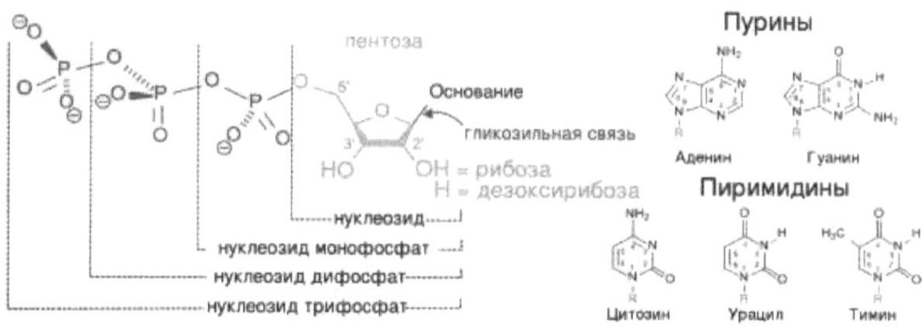

Figure 1.3. Nitrogenous bases (purines on top, pyrimidines on the bottom)

Title	nucleotides	Abbreviated title
as phosphates	as acids	
Lae nos 11 n - 5'-phosphate	5'-Adenylic acid	MMA
Guanosine-5'-phosphate	5'-Guaiylic acid	GMP
Cytidine-5'-phosphate	5'-Cytidylic acid	SMR
Uridine-5'-phosphate	5'-Uridylic acid	UMP
Dszoxyadsnosine 5-phosphate	5'-Dszoxyacetic acid	d MDA
De zoxy guanosi n-5'-phosphate	5'-Dszoxyguanillic acid	dGMP
Deoxycytidine-5'-phosphate	5'-Deoxycytidylic acid	dCMP
l'imidine-5'-phosphate	5'-Thymidylic acid	dTMP

Figure 1.4. Nucleotides that make up nucleic acids

In mild acid hydrolysis of nucleic acids, purine bases (adenine and guanine) are detached. If you increase the concentration of acid and heat the nucleic acids, the bonds between phosphates and sugars are broken randomly. The result is a mixture of free purine nucleotides, pyrimidine nucleosides bound to a sugar, nucleosides containing pyrimidine bases bound to a sugar and through the sugar to a phosphate group, which may be at the fifth or third

position of the sugar. Thus, a nucleic acid is a chain of nucleotides that are linked together by forming a bond between a phosphate and a sugar. In free nucleotides the phosphate' is attached to the fifth position, and in bound nucleotides to the third position. Such a chain is called the sugar-phosphate backbone or the sugar-phosphate framework. Normally, only the presence of one type of sugar in a biopolymer molecule is worth noting: either ribose or deoxyribose, but not their combination. These acids are called
RNA and DNA, respectively. In the case of DNA, the nitrogenous bases are adenine, thymine, guanine and cytosine (A, T, G, CD), while RNA has no thymine but uracil. DNA is unstable spatially but stable in environments, unlike RNA. RNA breaks down in an alkaline environment due to the presence of a hydroxyl at the third position, resulting in a rearrangement of the adjacent phosphate and consequent breaking of the bond with the previous sugar. The main habitat of DNA is the nucleus and RNA is the cytoplasm, hence they fulfil different functions. These functions were assumed to be related to active growth, because the RNA content increased if the cells were in the stage of active growth (Fig. 1.5).

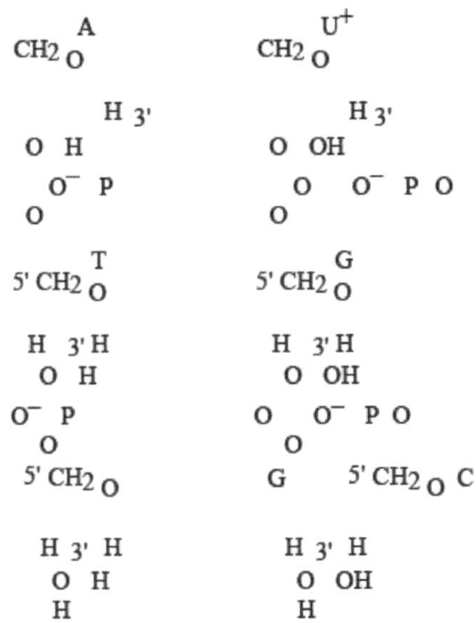

Figure 1.5: DNA fragment (left) and RNA fragment (right)

2. Structure of DNA, history of its discovery

Tetranucleotide hypothesis

In Figure 2.1 you can see how nucleic acids are organised chemically. Here is a fragment of a nucleic acid with all the atoms written out in detail. There is a 3'-end and a 5'-end at the other end, there is also a phosphodiester grouping where the phosphate is bonded to the hydroxyls of different riboses
- in the fifth and third positions, thus creating a chain of alternating sugar and phosphate, with nitrogenous bases sticking out to the side (ri[p. 2.1])

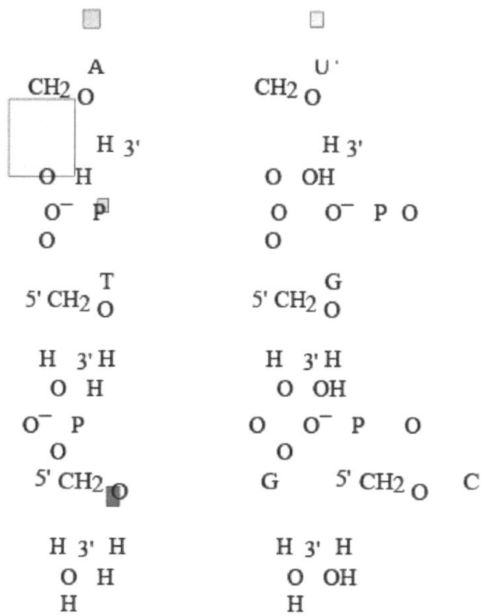

Puc. 2.1. Nucleic acid fragment

Such structures are irregular polymers, just like proteins (nucleotide alternation can be any). But this was not initially guessed, because these compounds are few in cells, it is difficult to obtain them and their analysis was not very accurate before. In parallel with the arrangement of elements, the number of bases was also determined. The research was conducted on DNA isolated from thymus, and it contains the number of nucleotides approximately equal, with a slight predominance of adenine and thymine over cytosine and guanine, but initially thought that the number of all nitrogenous bases is equal. Thus, the tetranucleotide hypothesis was put forward: there is a certain monomer built from 4 different nucleotides, i.e.

quadruplets alternate in the chain. This kind of structure could not be fundamentally different from structural polysaccharides. Therefore, initially there was an opinion that it was either structural, or spare components, or waste, and this inhibited the study of nucleic acids, because it was difficult to work with them, and why - it was not clear at that time.

The phenomenon of genetic transformation, G riffit's experiments

The solution to the problem came from a slightly different area. The English biologist Griffith found that if a culture of pneumococci, which possess a protective capsule of a polysaccharide nature that protects them from the host immune system, is grown for a long time on rich medium, abnormal colonies of cells appear that are hard to detect under the microscope but easy to detect on solid medium. As opposed to hemispherical smooth shiny shiny appeared more flat, wrinkled, matt. And if these colonies are sown on liquid medium, then all growing cells will have the same properties (inherited changes). A mutation has occurred - the formed cells have no capsule (Fig. 2.2).

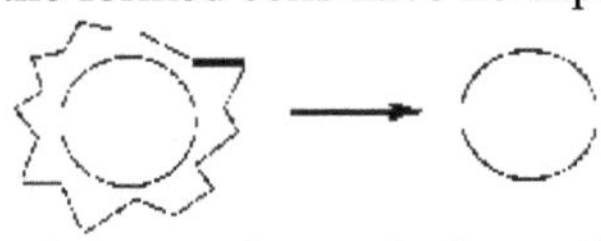

Figure 2.2. Scheme of capsule-free cell formation

As it turned out, such cells grow only in an artificial environment. Since when such a culture is injected into a mouse, the mouse does not die (the capsule-free strain turned out to be non-pathogenic). Moreover, both cultures were killed when heated to 60^0C. Injection of both strains into the mouse after heating did not result in mouse death. Injection of heated capsule-free and unheated capsule-free strains into the mouse simultaneously resulted in mouse death. That is, the living capsule-free cells gained some properties from the dead capsule cells. By Koch's triad, the capsule-free cells gave rise to capsule cells. Dead cells transferred properties to living cells - genetic transformation (Fig. 2.3). Now this phenomenon is widely used to produce GMOs.

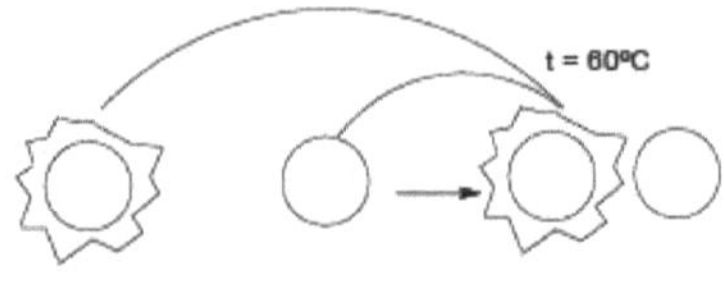

Figure 2.3 Griffith's experiments

Avery's experiments

In the forties, this effect interested the American chemist Avery, who was engaged in the chemical study of polysaccharides. He was interested in how

polysaccharides were synthesised in bacteria. He hypothesised that if there is no capsule and it never forms, and in another case there is a capsule but the cell is dead, then the capsule serves as a "seed" for the synthesis of a new capsule. That is, if the capsule-less has no primer, then no capsule is formed. He took dead cells, isolated purified polysaccharide from them and added it to the capsule-free cells, but found nothing. He then surmised that it was a matter of changing the heredity of the cell. So he isolated a fraction of proteins from the killed cells and added the fraction to the capsule-less cells - no transformation occurred. Then he surmised that lipids were the seed, isolated the lipid fraction and was also unsuccessful. So he decided that it was necessary to study nucleic acids. He isolated DNA from bacteria, added it to the capsule-free strains, and the mice began to die at the
by injecting them with these strains. Thus, if DNA is isolated from the capsular strains, added to the capsule-free strains and injected into mice, the mice die and, in accordance with Koch's postulates, the same capsular cells are isolated from here. That is, DNA carries genetic properties (Figure 1.4).

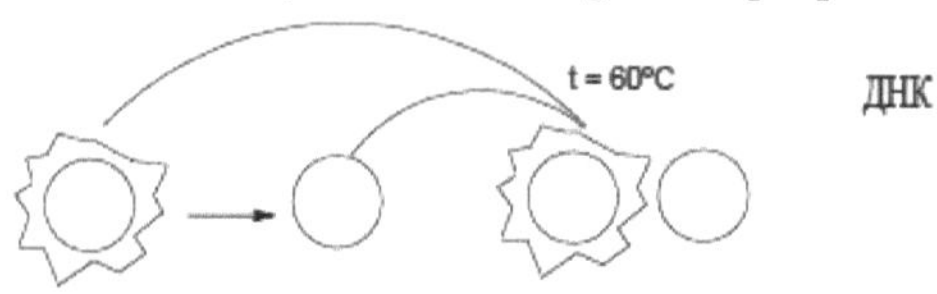

Figure 1.4 Avery's experiments with DNA

But the DNA was bound to proteins, as there was no possibility of high purification at that time. So Avery changed his experiment a little. He took an enzyme that breaks down DNA (DNAase was

commercially available.) Alien DNA is the worst enemy of all living organisms. And so is RNA. So all living organisms defend themselves against foreign nucleic acids by producing enzymes that destroy them.
Until recently, working with RNA was difficult because all living organisms are coated with a layer of RNase, and we are included. The first good work with RNA came after an experiment was presented in Nature in 1957. RNA from a virus was isolated and its absorbance in UV was taken, then acid was added to one part of this solution and RNA was observed to precipitate, and in the second part the experimenter dipped his finger in and stirred the solution with it for a couple of minutes. After incubation, he added acid to this part of the solution and observed no precipitation of RNA - all the RNA had decayed and was not precipitated. Entry of foreign DNA through the digestive tract is considered the most dangerous. Therefore, the human body has a very powerful and very resistant enzyme that breaks down these nucleic acids. Our pancreas produces DNAase and RNase, and RNase retains its activity even if the gastric juice is boiled. When 5% sulphuric acid is added, all proteins precipitate, but DNase and RNase remain in solution. One of the American firms at that time started to produce DNAase and RNase.
Avery took DNA from the capsule strains or the cells themselves, treated them with DNAase and mixed them with the capsule-free cells, and then injected them into the mice. The mice remained alive. The DNAase treatment removed the effect completely. Since nothing else but DNAase was added to the solution, it was proved that DNA is responsible for the process of transferring genetic properties (Fig. 2.5)

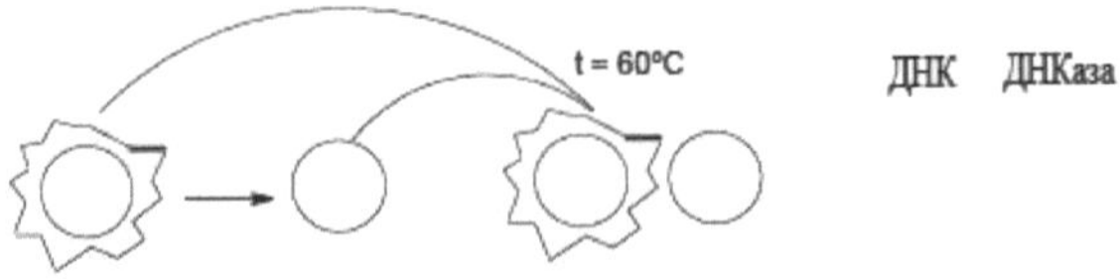

Figure 2.5. Avery's experiments with DNAase

Other experiments with the destruction of nucleic acids have also been carried out in parallel. In Austria, experiments were conducted on the tobacco mosaic virus. The virus is simple: it consists of one RNA molecule, which is covered by a protein shell. If you put this virus in a solution with pH 10, the RNA and protein will separate, and after adding acetic acid to a certain concentration, the RNA will precipitate, proteins remain in solution, so they can be separated. The experiment consisted of separating protein and RNA and infecting a tobacco plant, but at first nothing came out because the RNA was applied to the surface of the plant with a finger. The RNase on the surface of the finger destroyed the RNA of the virus. Therefore, the experiment was carried out by applying RNA to the surface of the leaves of the plant using a glass rod, which had been burned and treated with alcohol. The RNA then caused infection, and full-fledged viruses were obtained from the infected leaves. The protein did not cause infection. Then they took two different strains of tobacco mosaic virus, the proteins of which differed in charge: one had a positive charge, the other - negative. They separated the proteins and RNA, and then assembled them crosswise: RNA1+protein2 and RNA2+protein1. It turned out that the infected particles gave in the offspring only those particles to which the RNA corresponded, that is, when infected with RNA1+protein2, all the descendant viruses had protein of the first type. In this case, it turned out that in the tobacco mosaic virus, RNA is the carrier of genetic information.

Other work has been done on T-bacteriophages. The hereditary material for them is DNA. During infection, the phage injects DNA into the cell. In the work, the viral protein was methylated with radioactive sulphur and the viral DNA was methylated with radioactive phosphorus. The bacteria were infected with the virus and shaken. All the phosphate was transferred inside the cells.

In viruses, the hereditary material can be DNA and RNA, and in cellular organisms, DNA.

Chargaff's experiments with the ratio of nitrogenous bases

Nitrogenous bases are known to absorb well in the UV visible region (260nm). In the post-war years, the production of quartz, which is required for UV optics, was established. Also in the late 40s, the method of chromatography was rediscovered. This was related to the separation of amino acids, important in terms of determining the nutritional value of protein. Chromatography is a universal method that allows the separation of different substances that are close in structure (for example, the monosaccharides glucose and fructose). The method of paper chromatography for the separation of bases was applied by the American chemist E. Chargaff, who, in addition, developed a good method of hydrolysis

DNA. He took a preparation of DNA and hydrolysed it with 72% perchloric acid. Under these conditions, the acid acts as an oxidising agent on the carbohydrates, so the sugar part was completely destroyed. A solution with phosphate and nitrogenous base was obtained. By neutralising and filtering the solution, Chargaff obtained a solution containing only the nitrogenous bases. He then separated them using paper chromatography. On a strip of paper he would apply the preparation, pass the solvent through, after drying, place it in a mercury lamp (UV source) and observe dark spots on a white background, and, regardless of the system, each time he observed four sequences - these were adenine, guanine, cytosine and thymine. After chromatography, each piece of the base chromatogram was cut into thin pieces and placed in an acid solution to dissolve the base to then take a UV spectrum. In this way, Chargaff proved that the ratio of bases is not the same as previously thought. He also found that the nucleotide composition of DNA is species-specific - the first structural indication that the molecules serve as hereditary matter. The most important thing Chargaff discovered: the amount of adenine equals the amount of thymine, and the amount of guanine equals the amount of cytosine. It turned out that adenine and guanine are purines, and thymine and cytosine are pyrimidines, and the amount of purines equals the amount of pyrimidines. Adenine and cytosine have an amino group in the sixth position, and thymine and guanine have a carbonyl group. Thus, Chargaff concluded that 6-amino groups are equal to 6-keto groups (Fig. 2.6).

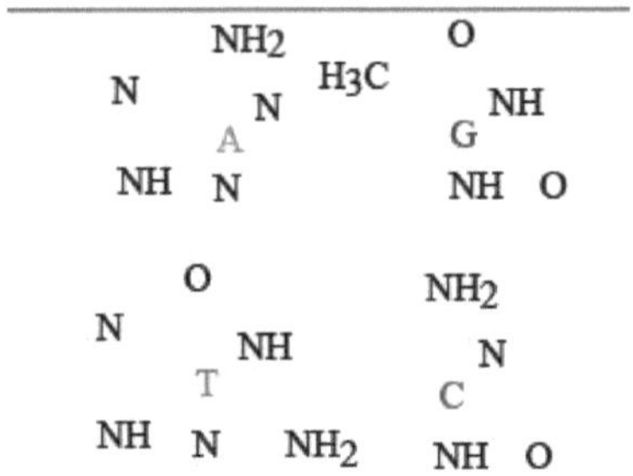

Figure 2.6 Nitrogenous bases

He surmised that if the ratios of bases are such that the concentration of one is always equal to the concentration of the other, it means that they are somehow related. This formed the basis of the following work. It can also be seen that the amino group is the hydrogen bond donor and the carbonyl group is the acceptor. But at that time the role of hydrogen bonds was not yet fully understood.

The next step was to attempt X-ray structure analysis (PCA) of DNA. This was carried out by Wilkinson's laboratory, in particular by P. Franklin. It is not possible to make crystals from DNA as DNA is a strand, not a knot. Therefore, the approach used for polysaccharides - with orientated films - was used.

Contributions of Watson and Crick. DNA geometry model, complementarity

Watson and Crick summarised all the data obtained. They became interested in the genetics of microorganisms, from which it became clear that fission is always binary. After a visit to Wilkinson, Crick, after looking at a drawing of X-ray structural analysis, immediately concluded that it is a spiral and derived the equation of light scattering of Rentegnian rays on spiral structures. They also studied Chargaff's work. They then decided that they needed to look at the DNA model problem. The PCA data did not match the nucleotide sizes well (the helix obtained in the figure was larger). At the same time, Polling suggested that the model would be similar to the α-helix, only the framework twisted into the helix would be ribose phosphate, with the bases sticking out. Watson and Crick ordered models of nitrogenous bases, phosphates, and sugars from a workshop and began assembling them into chains. It turned out that guanine could have an amino group opposite the corresponding keto group of cytosine. The distances and angles are such that a hydrogen bond is strongest when the hydrogen lies in a straight line with two electronegative atoms. The same is true when considering the pair of adenine and thymine. The first pair forms three hydrogen bonds, while the second pair forms two. Both pairs occupy the same space. Sterically identical pairs are formed. The distances between the first carbons of deoxyribose are equal and the bond angles are equal. The property of A-T and G-C pairs to have the same size is called complementarity, i.e. the pairs complement each other to some standard structure, and when there are complementary bases along the entire length, two chains - two polynucleotides - are said to be complementary to each other (Fig. 2.7).

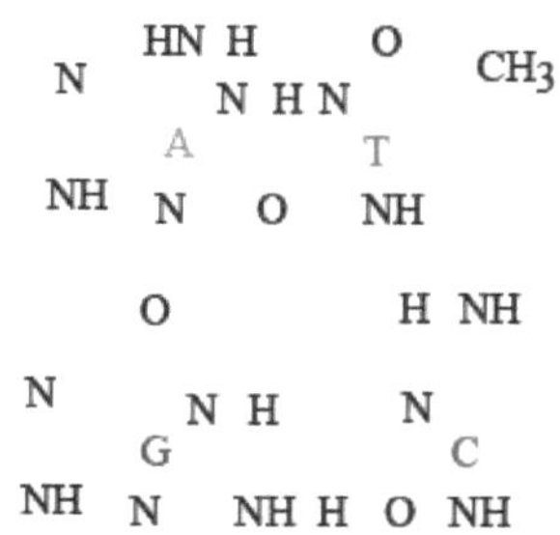

Figure 2.7. Complementarity of nitrogenous bases

Antiparallelism of chains. Grooves, significance. Parameters of the helix

An individual pair forming 2 or 3 bonds will not be bound very tightly. Watson and Crick built a model for several pairs alternating after each other in a chain: pair A-T, pair G-C, each of the same size. If these pairs in two chains are opposite each other, the spacing between the chains will be the same along the entire length, i.e., a regular structure will result. The regularity was indicated by the PCA data. A difficulty arose in the formation of such pairs. In addition to the distance between the chains, determined by the length of the rings, there is also the thickness of the rings. It's quite small. Therefore, they placed the bases on top of each other, due to the fact that the ribose and phosphate are tilted - a helix was formed. The bases form a continuous "stack" with a kind of twist. The planes of the bases are parallel to each other and perpendicular to the axis of the helix. In reality, the structure looks like nitrogenous bases sticking together in the centre and the ribose-phosphate framework wound around. The formation of pairs is possible if there is a certain relative movement, a certain arrangement in the chains. The bases linked by hydrogen bonds are rotated 180° relative to each other, i.e., if guanine has a hexagon as the bottom surface, then the hexagon linked to it by cytosine is the top hexagon, if we consider the direction of the chain. This leads to an important property of the double helix of nucleic acids - they are antiparallel (Figure 2.8). It is important to note that the angular distance with the framework on one side is smaller than on the other: the two sides of the planes are non-equivalent and differ by about half. If we translate this to the whole chain, there are two recesses - large and small grooves, which, alternating large and small, form the same helix as the ribose phosphate framework. Into these grooves extend the lateral surfaces of the nitrogenous bases. And having certain recognising systems (e.g. specific proteins), it is possible to recognise the sequence of bases at this location. An alpha helix fits well in a large groove, and a beta structure fits well in a small groove.

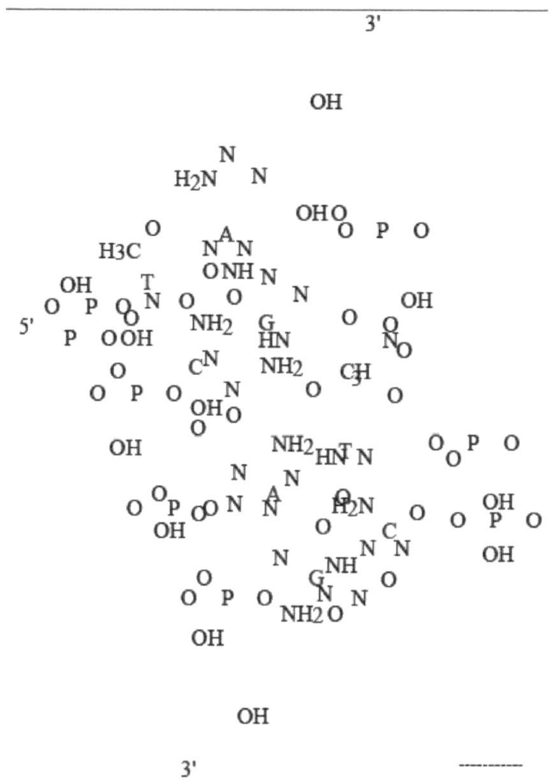

Figure 2.8. DNA structure

Crick later found out that quite strong van der Waals interactions occur between base pairs. This causes the twisted structure to become stable. There are a huge number of hydrogen bonds in the DNA structure (for 10 bases there are 25-30 hydrogen bonds in each chain). The phenomenon of co-operativity is observed here: when a long chain has cross-links all along its length cross-linking it, breaking one bond does not lead to a change in the structure, if this break is reversible. Therefore, such a structure is stable over a wide range of temperatures (up to 55-60°C the DNA structure is stable). Therefore, in Griffith's experiments pneumococci, which were not very stable, were not accidentally important; they died at 60°C, when the DNA was still preserved, because in this case the death was associated precisely with the disruption of the membrane.

3. DNA forms. The structure of chromosomes

A and B forms

The parameters obtained from the radiographs differed from the DNA model proposed by Watson and Crick. This was explained by the fact that there are several helical forms of DNA, and the transition between them is associated with certain external conditions. When the hydration of the molecule decreases, there is a form derived from the PCA data. This was a consequence of drying the DNA films for the necessity of the experiment. At humidity levels below 70%, an A-B transition of the DNA occurs. In the A-form, the bases are tilted to the helix axis with an angle of 70°. The angle between the ribose and phosphate changes, and in this form the grooves actually cease to exist. This A-form is most convenient for interaction with other molecules because of the greater accessibility of the groups enclosed in the grooves. The B-form exists in the living cell. The A-form can be stable when the sugar is not deoxyribose but ribose (the second position contains an OH-group. Then the hydroxyl hydrogen is brought close to the second chain phosphate in the next nucleotide, and a hydrogen bond is formed between them. Therefore, RNA is always present in the form of an A-helix (Fig. 3.1).

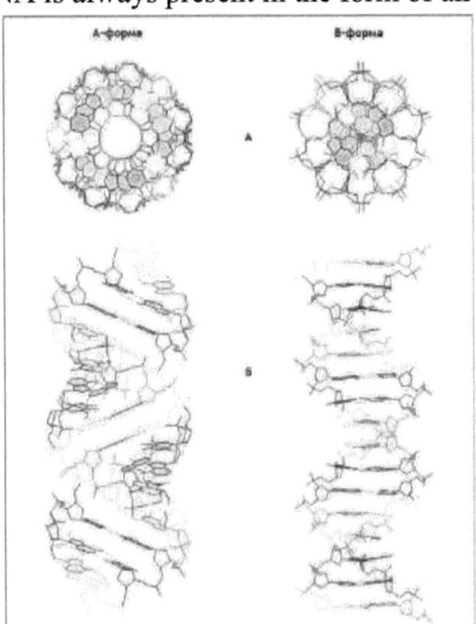

Fig. 3.1 Structures of A- and B-forms of DNA, A - end view, B - side view. Different nucleotides are shown in colours.

Z-form DNA. Comparison of all forms

It was necessary to confirm the existence of the DNA form according to the model of Watson and Crick. First they learnt how to obtain dinucleotides, then, modifying the technique, they were cross-linked. Two nucleotides were joined with guanyl and cytidyl bases, obtaining dinucleotide G-C, from which, by activation, a subsequent cross-linking of these dinucleotides of 5 doubles was obtained: GCGCGCGCGCGCGC. When the hexanucleotide is placed in solution, it forms a double helix structure because this chain is self-complementary. By selecting certain conditions, this small chain could be crystallised and X-rays were obtained of almost complete DNA, only with two bases. By then it was known that other transitional forms existed between A and B. The crystallographic data showed a left-handed helix, while the A- and B- and the transitional forms were right-handed helixes. In addition, the helix was jagged, zigzagging rather than smooth, so this form was called the Z-form. In this shape, phosphates are reversed every two nucleotides. Ribose is in the same conformation as in the B-form if it has a pyrimidine base (cytosine), and in the characteristic A-form if it has a purine base (guanine). That is, the residues seem to alternate in conformation. They also alternate not only in one chain, but also in two chains relative to each other: opposite to the A-form ribose (2'-endo) there is always in the second chain ribose in 3'-endo form. It is important to note that in this case there is also a rotation of nitrogenous bases. Therefore, the repeating unit in the Z-form is not a pair of nucleotides, but two pairs (Fig. 3.2).

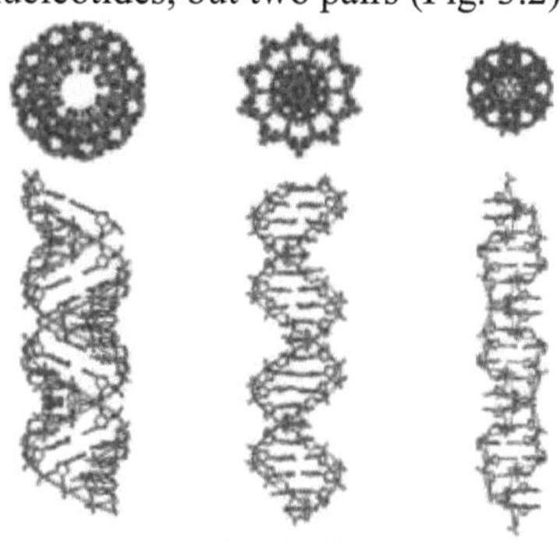

A B Z

Figure 3.2. DNA forms. The Z-form is formed at sites in the B-form where purines alternate with pyrimidines or in repeats containing methylated cytosine

In addition, the Z-form has characteristic spectral signals. This form is extremely scarce in living cells. However, from a functional point of view, specific proteins recognise the Z-form of DNA, but do not recognise the A- and B-forms. Antibodies have also been produced against the Z-form. Animals were injected with Z-DNA for a period of time into their blood, and antibodies were produced. The antibodies were isolated and chemically fluorescent labelled. Such fluorescent-labelled antibodies began to be used to stain cells. Those antibodies that found their antigen and bound, which in this case was Z-DNA, were supposed to stay and glow. It turned out that strictly defined areas of chromatin in the cell glow. That is, in certain, strictly defined areas of DNA in bacteria and chromatin eukaryotes DNA can be present in the Z-form. Further it was found out that DNA in any site can fold into Z-form, where purine-pyrimidine-pyrimidine-purine-pyrimidine alternate correctly...

H-form

There are several double-helical forms of DNA and under the action of proteins the standard structure of the B-form can be deformed, non-standard transitional conformations between A and B appear, whence it follows that not all 100% of DNA is in the ideal B-form. Speaking of DNA forms, we should mention the existence of one more form. This form is stable only in acidic pH solutions, so it is called the H-form ("ash"). The H-form is formed by two identical sections of DNA, i.e. sufficiently strict repeats, usually under slightly acidic conditions and only when one chain has only purine bases and the other chain has only pyrimidine bases.

The purine base forms one pair in the usual way through hydrogen bonding of amino and keto groups with the pyrimidine base. The keto or amino group present at the purine above can interact with the upstream purine also forming a hydrogen bond pair. This is well observed for the A-T pair because adenine has an amino group on top and thymine has two keto groups. It turns out that the keto groups of the two thymines will interact with the one amino group of adenine. For such interaction to occur for cytosine, it is necessary to attach one more additional hydrogen to nitrogen, which is possible only in acidic pH. In doing so, the cytosine is charged, but can form a hydrogen bond. Therefore, such structures are stable only in an acidic environment. Thus, a three-helix structure is formed and it is connected, as a rule, with the regulation of of gene work. That is, the formation of a single-stranded site makes it possible to start reading information from this place (Fig. 3.3).

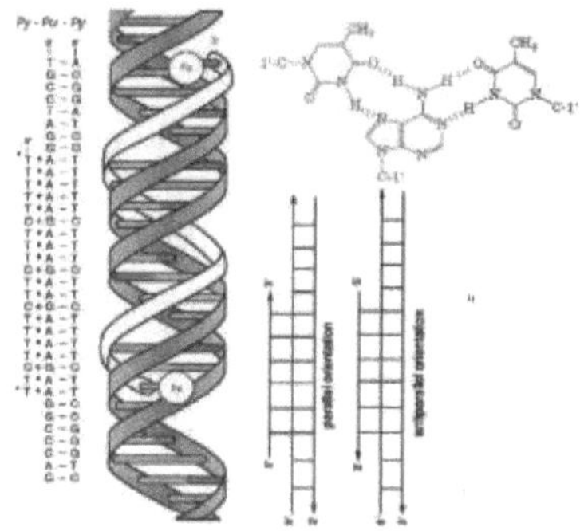

Figure 3.3. H-form of DNA (triple helix)

Superspiralisation, functions. Topoisomerases, classes. DNA gyrase

Since it is impossible to isolate the entire DNA molecule, we work with fragments of it. However, small DNA molecules are found in viruses. In the stages of working with viruses, it turned out that they have different genetic material. It has been shown that RNA viruses (e.g. tobacco mosaic virus) can infect a host plant without proteins, by introducing only RNA. For DNA viruses, the following pattern is observed: if DNA is isolated from the virus, separated from the proteins and centrifuged, two zones are formed. At first it was assumed that this was due to protein impurity. Then they purified the DNA from the proteins beforehand, but they also got two zones. Further each of these zones was centrifuged - the lower zone of the original is also divided into two, and the upper - so remains one. From electron microscopy (EM) results, it appeared that one zone had ring molecules. It was guessed about ring molecules earlier, because viruses have genes in DNA, which can be studied by genetic methods, a genetic map can be built. The map was circular, if you put aside the distance between different genes, they would just lie on the ring. In the second zone, however, there were molecules that were actually a ring, twisted into a rigid stick, which with the same mass is smaller in size, so it settles faster (less resistance). When standing, DNA gradually changes from a twisted form to a relaxed form - an open ring. There is a break before the twist, followed by a rejoining after the twist. In fact, it is the removal or addition of one of the turns of the helix. This phenomenon has been called DNA superhelicalisation.

It turned out that in those organisms in which DNA is circular, it is always superhelicalised. Moreover, superhelicalisation has a sign: it can be twisted in the same direction as the original double helix (positive superhelicalisation, twisting to the right). Negatively superhelicalised DNA occurs if there is DNA containing 1000 nucleotide pairs, i.e. there are 100 turns in the B-form and it is necessary to make one less turn - 99, but in order for the form to be stable, the molecule will twist in the same place, just in the other direction. DNA in cells is almost always negatively supercoiled. This came as a surprise. It was assumed that superhelicalisation was a special process. And, on the one hand, by looking for mechanisms that prevent positive superhelicalisation when replication is going on, and, on the other hand, by investigating those mechanisms that lead to normal negative superhelicalisation, enzymes were discovered. Topoisomerases are enzymes that do not carry out any reaction. It's a paradoxical situation: an enzyme, but its operation does not lead to any chemical change. The point of these enzymes is that they twist and untwist DNA. That is, they make a break in the DNA, turn the strands relative to each other, and cross-link them. Since as a result the same phosphodiester bond is formed, there is no chemical reaction in principle, but there is mechanical work. And if there is mechanical work, then there must be a source of energy. Where does it come from? Well, it turns out that topoisomerases are divided into into two classes according to as to what kind of energy energy they use. And at the same time by the mechanism they realise. Topoisomerases of the first class reset excess turns: positive or negative. That is, a ring is formed from the superhelix. The enzyme binds to practically any section of DNA. As a result of superhelicalisation, a certain section is unraveled. this conformational transition requires some kind of energy boost for the individual parts, and this happens only when there is tension in the DNA itself. That is, the structure - double-helical DNA and a single strand next to it - is only possible when there is a supercoil at that location. As the supercoil is dropped, the energy disappears and this process will not happen a second time. It will not go in the opposite direction either, because energy input is required, and there is no energy source. Topoisomerases of the first class are quite numerous and have an important role in replication: when positive supercoiling occurs during rapid DNA unwinding, they always reset this supercoiling (Fig. 3.4).

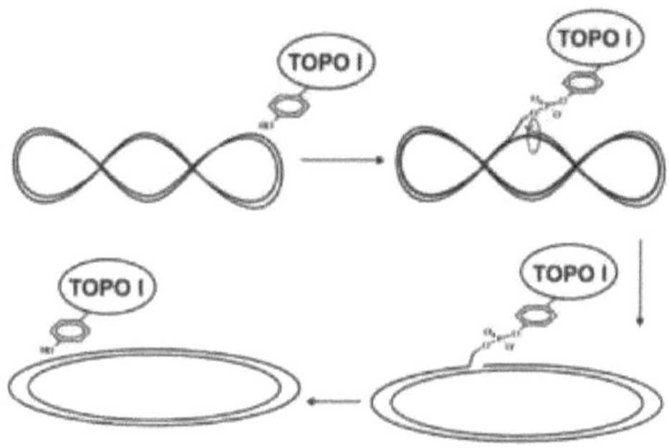

Figure 3.4. Topoisomerase I

The second class of topoisomerases uses an external energy source in the form of an ATP molecule. At the same time, it works not with single-stranded but double-stranded DNA - it makes a break in the double helix. And in order for this to lead to some kind of rearrangement, it doesn't just bind to the double helix. When superhelicalisation takes place, an enzyme "sits" at the site of the cross-over. The enzyme is made up of two types of subunits of two copies each. The free enzyme usually has a conformation where there is some kind of open gap where the first strand of DNA goes in one direction. After. DNA has bound, there is a mid-strand closure, a second strand similarly comes in the other direction, makes changes to the structure of the protein, and the resulting active centre cuts the two strands of DNA side by side, forming a double-stranded break. At the site of the break, the subunits open and the second strand of DNA comes out, passing through. The enzyme parts close together and the first DNA strand joins,

the upper passage opens again for the next DNA strand (Fig. 3.5).

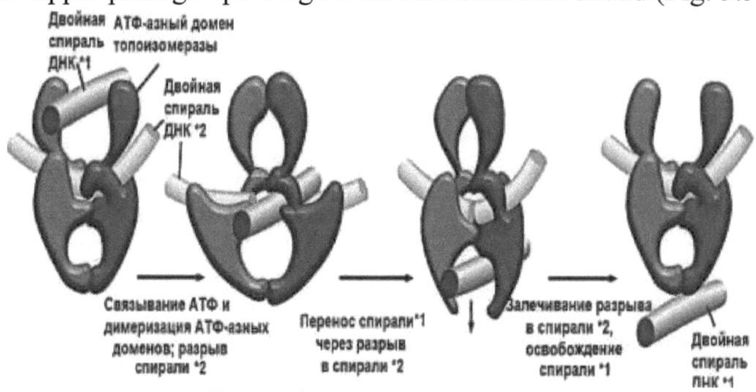

Figure 3.5. Topoisomerase II

There is no actual DNA strand break, as the bond is not broken, but transferred. The enzyme has a tyrosine residue in its active centre and works in such a way that the phosphodiester bond is broken at the binding site with the hydroxyl of ribose and transferred to the hydroxyl of the tyrosine residue. As a result, the ends are covalently cross-linked to the protein and a free 3'-end is formed. The energy costs are the same for bond breaking and bond formation, which is why topoisomerases of the first class have no energy costs. In the case of the second type of topoisomerase, there are two such centres, and each of them binds only one of the two chains that were formed after the break - the 5'-end, since the 3'-end does not carry phosphate, it is hydroxylated. So each of the double-stranded molecules is bound by only one strand. But since they are complementarily bound together, this causes both the first and second strand to be fixed together.

This cycle involves ATP as a source of energy. ATP binds to the molecule at one of the initial steps: when the second strand is attached, ATP binds in parallel, and then the bottom channel opens to allow the DNA to be "dragged" from the top to the bottom. When the strand passes downwards, ATP hydrolysis occurs and the system returns to its previous state. By hydrolysing ATP (an irreversible process) a large amount of energy is created, so the whole cycle becomes irreversible. The ATP energy is not spent on any real work, it only serves to move ATP out of the protein, all the work done before that is done by the energy of thermal motion. ATP causes a conformational transition in binding, but it provides no energy in doing so. So class II topoisomerases make double-stranded breaks, and class I make single-stranded breaks.

Topoisomerases are involved in almost every process in which DNA is involved. It turns out that in almost every case where something happens on DNA, it has to be in a negatively supercoiled form. Because in order for unraveling to happen more easily. negative superhelicalisation. Negative superspiralisation is a lack of coils. There are special proteins in cells that stabilise the unbraided state. In the case of negative superhelicalisation, it statistically occurs and is fixed by this protein that binds there. That way, it becomes possible to create a specific single-stranded region where the DNA is readable. Such processes are needed to start the doubling of DNA and to read from it in the form of RNA. Therefore, if anything in the cell disrupts the negative superhelicalisation of DNA, the cell will not be able to reproduce and grow because RNA synthesis will stop, new RNA will not be formed and the old RNA will decay.

Bacteria have a special form of class two DNA topoisomerase, DNA gyrase. This protein is the target of quite a large number of antibiotics. It turned out that a large number of antibiotics simply disrupt DNA superhelicalisation. That is, practically nothing changes under the action of such antibiotics. They simply reduce the replication rate because there is no negative superhelicalisation.

Eukaryotes also have topoisomerases II, but they are not affected by antibiotics of this type (particularly chloro-quinone antibiotics). Superhelicalisation is fixed if there are covalently closed rings. So for eukaryotes, for DNA to replicate, and for RNA to be synthesised on it, negative supercoiling is also needed, and this is done by topoisomerases embedded in chromatin, and because they have to hold loops of relatively small size, there are many of them, and the amount of topoisomerase content is higher in eukaryotes than in prokaryotes. This negative superhelicalisation is in certain cases compensated for by the positive charge of other sites. That is, in general, DNA appears to be either unhelicalised or weakly helicalised, but in fact there are many supercoils, it is just that they are positive at one site and negative at others. In particular, in eukaryotes, a certain positive superhelicalisation is formed when DNA is bound in chromatin structures, i.e. where DNA is as if wound onto filaments and protein scaffolds.

DNA stacking. Chromatin, chromosome structure

There is a certain paradox of DNA size. If you take E. coli, it has one DNA molecule in it normally (if it is not replicating), one that is about 1 mm long with a cell size of 2 microns long and 0.8 microns in diameter. It turned out that this is possible because DNA is not just randomly arranged, but there are special proteins and RNAs that fix its specific regions with the formation of loops. In fact, DNA domains are formed, which are comparable in size to the size of the cell, but each of them is directed in different directions and does not go outside the cell. In the human body there are 48 chromosomes, each containing its own molecule, and in total the size of DNA turns out to be 120 cm. All of this fits in a nucleus with a diameter of 4-5 microns. To solve the problem of tangling of coils in eukaryotes there is a special step-by-step stacking of DNA with the help of special proteins - chromatin is formed. The formation of chromatin begins with the formation of nucleosomes. Back in the late 19th-early 20th century, it was discovered that there was a group of positively charged proteins in the nucleus of eukaryotic cells. As the cells of the nucleus were studied, they learnt to isolate the cells, and from them they learnt to extract certain structures containing DNA, and it turned out that all of these proteins were part of the structures. These proteins, called histones, form complexes with DNA. It was originally thought that these were simply electrostatic complexes, meaning that the counterions were attracted to each other. But this did not explain some of the peculiarities of histones: there are a strictly defined number of histone species in the same living organisms, otherwise any sequence could be used. There are five kinds of histones: they differ in size and in the content of different amino acids. They are labelled by the letter H ("ash") with a corresponding numerical index according to the order in which they are arranged during the separation process on specific media: H1 H2 H2 H3 H4, then it turned out that H2 is not one but two different proteins, and they began to be called H2a and H2b. H1 mainly contains the amino acid lysine as a positive charge, the protein is larger than the others (25 kDa). H2a and H2b are close in size and contain approximately equal amounts of arginine and lysine; it was because they are close in size and charge that they initially could not be separated, while H3 and H4 are arginine, many times smaller in size than H1 and 1.5-2 times smaller than H2. It turned out that H3 and H4 are very conserved - in different species of organisms they are very close in amino acid sequence, not just in composition. Histones were isolated by hydrochloric acid extraction from the nucleus. When they were isolated, they turned out to be about the same mass as DNA. And the number of different forms of histones, if to recalculate on molar units, instead of counting by mass, it will appear that also approximately the same, it, in turn, indicated the formation of structures of constant composition - complexes. It turned out that when extracting the nucleus in a certain way, a complex of DNA with histones is formed, which in EM will look like DNA strands with spherical "beads" on them, such structures were called nucleosomes. There are two molecules of histones H2a, H2b, H3 and H4 per particle, and a histone octamer is formed - a dense, stable association (Fig. 3.6). The histones join together to form a single globule, and this structure is necessary for DNA to be wound onto it. Although the attachment of DNA to histones is independent of the nucleotide sequence, there is some advantage to some sequences, they are more often wound in this way, and the distance between nucleosomes is strictly constant. A permanent unit consisting of a nucleosome and a linker is obtained.

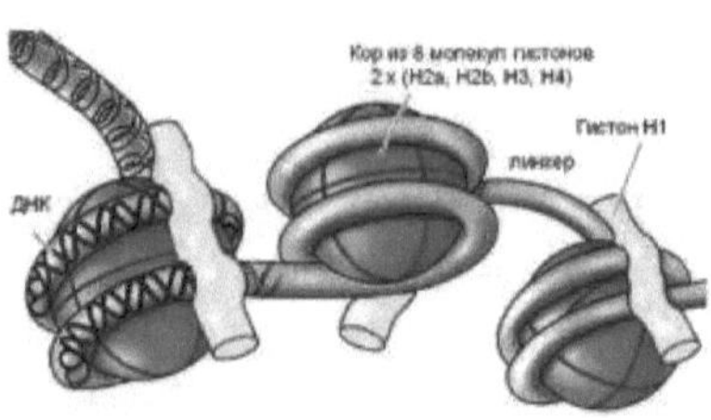

Figure 3.6. Structure of a nucleosome

The particles were discovered by gently treating chromatin with DNAases.

The section between nucleosomes is broken under the action of certain enzymes that cleave DNA, but if few enzymes are taken, the break occurs literally one or two bonds. Thus obtained these particles in almost native form, and even managed to crystallise and figure out the spatial structure of the nucleosome. Through this it was discovered that histones have protruding tails, histones H2a and H2b consist of two parts: a dense globular part, which is part of the nucleosome and a tail, the tail contains positively charged amino acids. Therefore, when DNA is coiled, there can be two options: if this tail is modified in some way, and the charge in it is reduced, it remains hanging freely; if it is made positively charged (for example, by methylating lysines), it will coil on top of the DNA and block it, this section will be closed and information cannot be read from it. So there are modifications to histones that make them repressors, or allow genes to be read out. Such a nucleosomal filament undergoes stacking, the filament and nucleosomes are folded into a fibril. It has different parameters depending on how it is produced. Nucleosomes interact with histone H1, which binds to the linker with linker sequence, and H1 interacts with each other and as if pulls these sites together. A kind of thicker strand is produced. In the third step, loops are formed (as in bacteria) due to fixations on the base (scaffold), which are proteins forming a long axis from which the loops branch off, there is topoisomerase-II at the attachment points, so each loop can be negatively superhelicalised. These loops then form rosettes, which, sticking together, twist into the next helix. When the rosettes stick together, chromosomes condense. Until then, DNA is in interphase (distributed throughout the nucleus), and further leads to the formation of compact chromosomal structures, which are convenient to transport, but inconvenient to read information from genes and replication. Therefore, replication occurs first, followed by condensation and divergence.

Different areas of DNA in a cell are compacted differently. There are areas that are always tightly packed and never work: heterochromatin - seen on histological preparations as more intensely stained. There are also highly loosened areas in which certain genes can occur, and a third very small part - genes that work at a given time - may even be devoid of nucleosomes (this applies to ribosomal DNA genes, nucleosomes have almost no time to assemble on them, because transcription takes place there all the time). The nucleosomes themselves do not interfere with the process of reading information, because the enzyme that does it knows how to bypass them, bending the DNA so that the nucleosomes move from the place that is being read to the place that has already been read. In this way, loop transfer makes it possible to read nucleosomal strands.

4. Bacterial DNA replication

The Meselson and Stahl experiments. Semi-conservative DNA synthesis

Watson and Crick suggested that the exact complementarity of two chains allows the cell to separate these chains, and then on each of them to complete the second chain according to the principle of complementarity.

The evidence soon came from two American researchers, Meselson and Stahl, who used the DNA centrifugation method to distinguish between old and new DNA. The main problem with polymer synthesis is that there are a large number of polymers in the cell and the new ones make up a smaller proportion, so it is difficult to identify the new protein or new RNA, DNA that has formed in the cell because of this. Meselson and Steel took the stable isotopes ^{15}N and ^{14}C, where the carbon is as part of glucose and the nitrogen is simply as an ammonium salt, as a medium for E. coli to grow. As a result of the bacteria repeatedly dividing, their DNA is synthesised new all the time, it will gradually all be made up of these heavy isotopes. The DNA was then isolated and compared with normal DNA by centrifugation in a density gradient of caesium chloride (Figure 4.1).

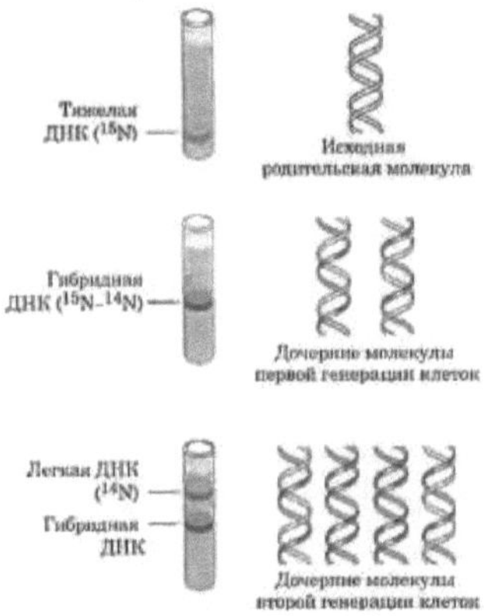

Figure 4.1. Meselson and Stahl's experience. Semi-conservative nature of replication

When DNA is in solution, it has phosphate groups around it holding counter ions. Cesium is a heavy ion, so cesium increases the density of DNA quite a lot, and if the DNA is clean, dry, it has a density of the order of 1.2 - 1.3, as well as proteins, then in such a gradient it behaves as a substance with a density of about 1.6-1.8. And this density range is well created by centrifugation of cesium chloride. This means that when DNA is placed in a solution of cesium chloride, the cesium will spread out, creating a gradient, and further down this gradient, the DNA, which has a larger size and lower diffusion rate, will begin to spread out, just like the cesium, under the influence of gravity - that is, it will settle down until it reaches its density zone. As a result, molecules that were originally distributed evenly throughout the centrifuge cup will end up in one narrow zone corresponding to their density. Meselson and Steel performed the same experiment by crossing cells grown in heavy medium to light medium (with the usual common isotopes ^{12}C and ^{14}N as carbon and nitrogen sources, respectively). After half an hour, the time required for cell division, DNA was also isolated and centrifuged in a caesium chloride gradient.

It is worth paying attention to two possible variants. If the new DNA were synthesised independently of the old one, as a separate molecule, as proteins, polysaccharides are synthesised, then the new DNA would all be light and the old one all heavy. In this case, light and heavy strands would appear. This is the case when new light DNA is synthesised from the original heavy DNA and the old heavy DNA remains. This mechanism is called conservative.

The second variant could be obtained when DNA was doubled according to the mechanism proposed by Watson and Crick: in this case, each of the chains of the initial heavy DNA serves to superstructure the second one, i.e. strands consisting of one heavy and the other - light halves are obtained, i.e. DNA should be of intermediate density.

There was also a third option, which was that DNA in some places was synthesised new on a given strand, and in others remained old - the so-called "mosaic mechanism". It was clear that it was unlikely, but in this case there was old DNA, as a result there was an old site, then a new site, then an old site, then a new site, and, accordingly, in the second strand it would be the other way round, or if we talk about two molecules, there could be different random distribution in different places. In this case, the same thing happened when both gave the same picture, which was reflected in the experiment.

Thus, when Meselson and Stahl isolated DNA after one replication cycle, they did not have two zones of light and heavy DNA, so the model of the conserved mechanism was wrong. What was left was to choose between the two. Then they grew the cells not one cycle, but two. In this case, two bands were formed: one zone the same the same as it was at the previous stage and a light zone appeared, the amount of which was about the same as in the hybrid one. It turned out that the two filaments diverged, and on the heavy and light synthesised additionally one light filament each - this meant that half of the molecules became of intermediate density, the other half - light. Thus, the semi-conservative mechanism of synthesis, which was proposed by

Watson and Crick, was proved.

Nucleoside triphosphates

This semi-conservative mechanism implies the existence of systems that unravel DNA, attach nucleotides to the old strand and check their complementarity, mechanisms that connect nucleotides, i.e. all this requires certain proteins, because it is proteins that carry out various processes in cells. So work began to discover these proteins and decipher them. The first thing that was identified was what nucleic acids are synthesised from. That matrix strand, which is there, must attach some form of nucleotide, and it turned out that such a form was nucleotide triphosphates (Fig. 4.2).

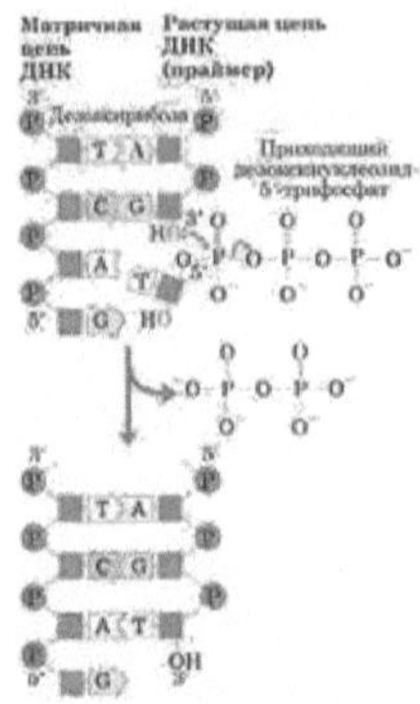

Figure 4.2. Matrix chain attaches nucleotide triphosphate at the 3'-end

Also, as in ATP, there can be all 4 bases that can be part of a nucleic acid. I.e. there can be thymine, guanine, adenine, cytosine - all in the form of triphosphates. i.e. ATP, GTP, CTP, TTP. They carry a large energy reserve, the phosphate groupings are connected by an anhydride bond, which is energetically very unfavourable - its hydrolysis results in the release of about 30 kJ/mol of energy. It is logical to assume that two phosphate groups connected by such a macroergic bond would be a good substrate. Accordingly, these triphosphates in case of DNA synthesis should have contained deoxyribose, such nucleotides were practically not found in cells. Much later it turned out that these nucleotides are not present in the cell in an amount greater than necessary for DNA synthesis. In eukaryotic cells, this enzyme does not work at all for almost the entire cell cycle, except for the S-phase. In bacteria, protein synthesis goes on continuously in fast-growing cultures, so it is always working, but it gives the products needed for DNA formation in just the right amount. It was therefore impossible to isolate nucleotides from the cell - they had to be chemically synthesised. But it was already clear that the right nucleotide had to be found first: for example, if adenine was in the chain, thymine had to join it.

DNA synthesis, DNA polymerase I. Okazaki fragments. Primase

In addition, it was shown that synthesis does not proceed along the entire chain, but pointwise. The site of doubling (nucleotide addition) moves along the original matrix strand of DNA together with the enzyme with gradual building up of the chain. That is, the strand grows from the end in a certain direction, and the enzyme first of all holds the end of the matrix, which should build up in the centre where the addition is taking place. The chains themselves are held together by inter-nucleotide bonds, but the end nucleotide can easily detach because it is not held together by its neighbours. The enzyme acts as a neighbour in the long chain and holds the place. Its task is to pick up the right nucleotide, this process is by random selection, as it is not shown that there are any special rearrangements in the enzyme depending on what is on the matrix chain, i.e. for any nucleotide the structure of the active centre is approximately the same. So it is the interaction that is important here, i.e. the wrong nucleotide can also join.

The enzyme, after a nucleotide has attached, checks the position of the phosphates of that nucleotide: they should end up next to the hydroxyl of the growing end. If everything is correct, they definitely end up in that location. It is now known that there is a rather universal set of amino acids in the active centre - a few acidic amino acids, such as asparagic acid, which with its carboxyls fixes several magnesium ions, and the latter binds to two carboxyls, finding itself clamped in a strictly defined place and already fixing phosphates and additionally weakens the bond between these two phosphate groupings. As a result, the bond between the phosphates is broken and the phosphate group attaches to the third position of the deoxyribose at the growing end. In this way there is an accretion of the chain one at a time.

It was necessary to isolate the enzyme to determine how it worked. From the presence of activity it was clear that triphosphates were needed. To study the activity of the enzyme it was necessary to have a matrix, nucleotides and to observe the formation of the polymer. Radioactive nucleotides were usually used for this purpose. The foundations of the work were laid by the eminent molecular biologist Arthur Kornberg Sr, who was the first to obtain in pure form the enzyme that synthesises DNA. He determined the activity by isolating DNA from an object, dissolving it, and adding ions to the solution (it was magnesium ions that were necessarily needed). He would then add nucleotides and an enzyme and watch the radioactivity that was in the nucleotides transfer to the acid-soluble fraction of the polymers. Since nucleic acids are acids after all, when added to a solution to a PH below 2, they do not dissociate and begin to stick together and precipitate. Accordingly, the newly synthesised chains precipitate out, and with them the radioactivity. The precipitate was then filtered and placed in a counter to determine the radioactivity. After purification, he obtained the enzyme in its pure form. The enzyme was quite active, and does require triphosphate for it to work. When we started to check its work on different matrices, an unexpected thing occurred: if we took not very homogeneous DNA preparation, for example, DNA isolated from some animal cells (cells of which are multichromosomal, respectively, many different DNA molecules, plus they are large, when isolated they break, fragments are obtained), the enzyme showed activity. When viral DNA (supercoiled ring DNA) was taken, it turned out that the enzyme did not work. It was logical to assume that it could not unravel it, but it turned out that if DNA is not supercoiled, but relaxed (as a result of a single break or a small number of breaks) - it is also a very bad matrix.

This led to experiments in enzymatic degradation of DNA. With the help of DNAase, which randomly breaks the chains somewhere in the middle. For example, they took animal pancreatic DNAase in very small amounts (in such quantities that there were a million DNA molecules per 1 enzyme molecule) and incubated it for a very short time. This caused a small number of breaks in the DNA, but because the enzyme only breaks one strand, these breaks were in one strand and the second strand remained intact. Thus, the DNA maintained the appearance of seemingly whole relaxed DNA, but in reality there were many cuts in the DNA. It turned out that the more DNAase was incubated (the more cuts were made), the more DNA polymerase synthesises on it. Then they began to examine the obtained DNA for the place where radioactive nucleotides are included. It turned out that they are included at the break point. That is, DNA polymerase lengthens the 3'-ends of the molecules formed as a result of the break and gradually attaches to them new nucleotides complementary to the second chain. The second strand, which was broken off at this point, is pushed back. Synthesis always starts with a break. The enzyme needs a 3'-end, that is, it is not able to start the synthesis of new molecules, but only lengthens the old ones (Fig. 4.3).

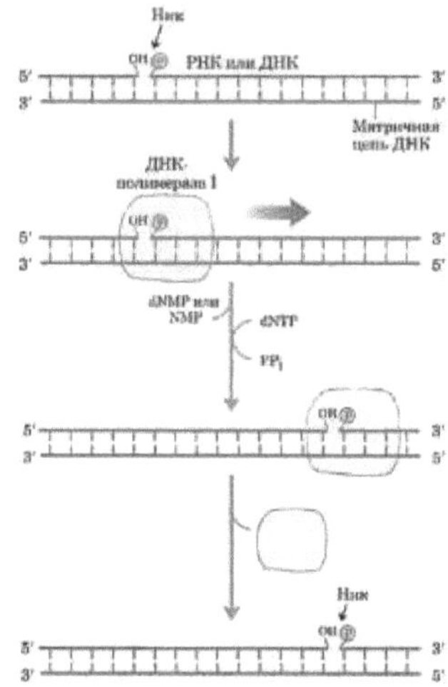

Figure 4.3. Action of DNA polymerase I enzyme after DNAase break realisation

Here it was found that as DNA polymerase works, initially positive supercoils accumulate in the DNA. That is, the strand does not unwind, but runs the coils lengthwise. Next, class one topoisomerases are switched on, making a break in one strand and pulling the second strand through it, dropping the coils due to energy taken from the DNA polymerase. In addition to topoisomerases, there is another class of enzymes involved in the spatial modification of DNA - helicases (helix unwinding enzymes). Helicase is constructed of 6 subunits. It braids the DNA, travelling along it, bringing down the coils, unravelling the strand. Further in front of it acts, when the coils accumulate, a class 1 topoisomerase. Thus, DNA polymerase receives already single-stranded DNA as a matrix. Special proteins (SSB-proteins) with affinity for single-stranded DNA are involved in this process; they sit on the unwound DNA and prevent the reverse twisting process. The process of the very beginning of replication remained an unsolved question.

Returning to DNA replication in real cells, attention was drawn to an earlier observation made by the Japanese scientist Okazaki: if we take unfavourable conditions or mutations in E. coli cells, DNA fragments accumulate. Placing this DNA in an alkaline gradient showed that the bulk of DNA, which is determined by ultraviolet light behaves like a large molecule, and newly synthesised DNA is represented by very short "pieces", i.e. there is not a continuous build-up of the end in a long molecule, but many short fragments are formed, called "Okazaki fragments". These fragments, which are several tens of nucleotides long, are formed and cross-linked after a certain time

(indicating their constant number) into larger molecules. Okazaki fragments are formed as a result of the fact that one of the strands of the synthesis is discontinuous, and for each fragment is formed its own inoculum.
From further experiments it appeared that with increasing retention time in the alkaline gradient the fragments shorten and DNA is stable in alkali. RNA is hydrolysed in alkali. It turned out that it was possible to leach part of the nucleotides from newly synthesised DNA, it turned out that this was the part where the fragment started (i.e. its 5'-end). An enzyme that synthesises RNA, which is an exception to nucleic acid polymerases, was found to work in complex with the DNA polymerase. This RNA polymerase that synthesised the beginning of the Okazaki fragments makes an error of 1 error per 10 nucleotides even more often. It turns out that this enzyme can start at any point, and it has no specific start sites. Frequent errors lead to the fact that the end of such RNA cannot be correctly positioned on the matrix, which makes joining the next nucleotide impossible, i.e. the incorrectly formed end pops out of the active centre. This enzyme, which synthesises *primers*, is called "primase". I.e. primase is a highly erroneous RNA polymerase that synthesises short fragments from any place in DNA (Fig. 4.4).

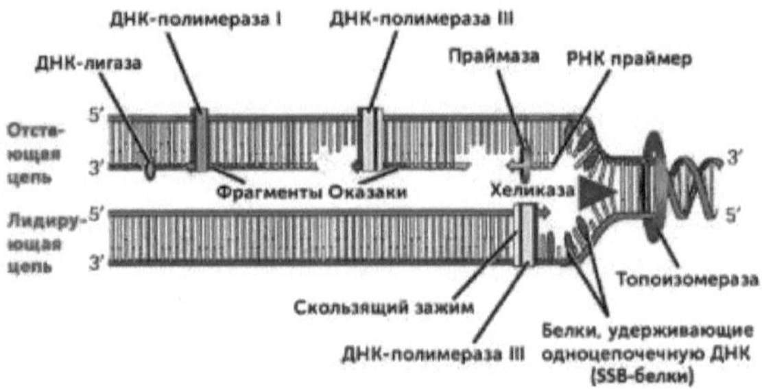

Figure 4.4. The role of primase within the DNA replication mechanism

Next, DNA polymerase comes into the process, which has another interesting property - corrective activity: if it accidentally attaches the wrong nucleotide or two, it destroys them itself by moving backwards. Once attached, DNA polymerase starts synthesising DNA. This process happens on one chain, on the second chain there is a problem due to the fact that nucleic acids are synthesised in only one direction - they attach to the hydroxyl in the third position, because the nucleotides that are in the cell have a phosphate in the fifth position. In the matrix, the situation is reversed because the chains in DNA are antiparallel. When an extended single-stranded region is formed and stabilised by special proteins, the enzyme synthesises the inoculum in the opposite direction at the end of the already unbraided region. And these resulting short stretches are Okazaki fragments.

Next, the resulting Okazaki fragments must be deprived of inoculum, as there should be no RNA in the mature enzyme, and the strands must be cross-linked together. For this purpose, it is necessary to consider some special properties of DNA polymerases. When working with nucleic acids, check the preparation for nuclease contamination.

Hydrolases (an enzyme of the nuclease class that hydrolyses compounds) can break the bond in several places: when it comes to DNA cleavage, there are two approaches - splitting off monomers from the end one at a time and breaking at some places in the middle. Such enzymes are called endo- and exohydrolases i.e. endonucleases break DNA in the middle and exonucleases break it from the ends. Endonucleases by a single break can turn the molecule into two halves, each of which is non-functional. The endonuclease may not stop at one break, but make a second and subsequent breaks. The two processes are very different in their speed - the exonuclease works much slower.

Nick-translation. DNA polymerases II, III

DNA polymerases have been shown to have exonuclease activity. There are two types of nucleases: 3'-exonuclease and 5'-exonuclease. Kornberg's DNA polymerase had both activities, i.e. it could cleave nucleotides from different ends (in artificial systems). It turned out that if it was subjected to proteolysis, the enzyme split into two parts. The protein thus consists of two globules linked by a region accessible enough for proteolysis. One globule had polymerase activity. And the other globule didn't. But this main part had only 3'-exonuclease activity. That is, this fragment (Klenow's fragment) had two activities: it could attach the next nucleotide to the 3'-hydroxyl if there was a matrix and a primer, but it could also detach the nucleotide from the 3'-end. It turned out that the molecule contains two active centres that can change places by rearrangement of the molecules. The second globule exhibits only 5'-nuclease activity and is able to remove nucleotides from the opposite end of the synthesis. This process, in which the polymerase cleaves off one strand and then completes a new one, is called shsk-translation (nick-breaking). At one time, this method was actively used to produce radioactively labelled DNA (Fig. 4.5).

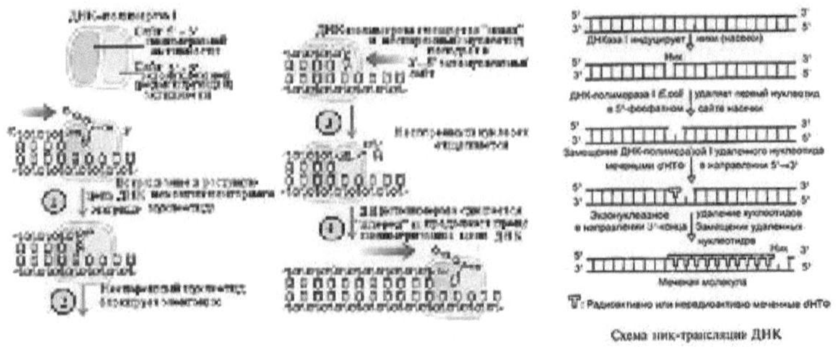

Figure 4.5. Nick-translation

Next, Kornberg found another enzyme, DNA polymerase II (the first enzyme he discovered was called DNA polymerase I or "Kornberg's enzyme"). DNA polymerase II is a completely different enzyme, smaller in size, lacking 5'-exonuclease activity, it was almost 100 times less active in cells.

It was further discovered that if we took mutants lacking polymerase I and II and looked at them by radioautography to determine the site of DNA synthesis, it turned out that the newly synthesised DNA appeared to be bound to membranes. A third form of DNA polymerase (DNA polymerase III) was isolated from the membrane fraction. It turned out that it actually carries out replication. In addition, this enzyme is characterised by the fact that it forms a stable complex with DNA helicase (i.e. unwinding is actually performed by one protein complex). This enzyme is highly unstable to high ionic strength, which is often used in protein purification. The enzyme is present in small amounts but has a very high activity, i.e. it is very fast. Another special property of this enzyme is its accuracy and low error rate.

Comparison of DNA polymerases. The structure of the enzyme complex DNA synthesis. Origin

It is interesting that specific polymerases of viruses make errors more often but also work faster, while eukaryotes make errors less often but work slower, so they need replication points. The third parameter that characterises polymerases is processivity. The processivity of I and II polymerases was somewhere around a few thousand nucleotides. The processivity of III polymerase could not be calculated. The solution to this problem was found when they studied how the enzyme is structured and some features of its work. (Fig. 4.6).

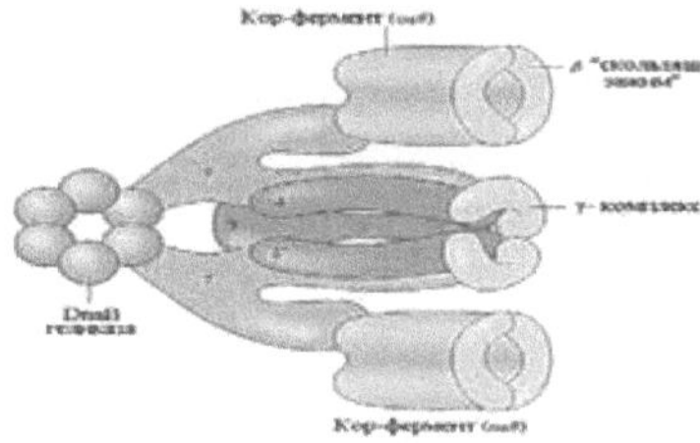

Figure 4.6. DNA polymerase III

The structure of the DNA synthesising enzyme complex includes several subunits that make up the complex: what is considered an independent enzyme, the helicase (shown as a hexamer that sits at the beginning, through which the DNA passes and unwinds), paired tau subunits that bind through the gamma subunit hold this complex and all the other subunits, delta and gamma subunits, they are structural, they define the assembly, a structure of three subunits (alpha, epsilon, tetta) that can be separated under certain conditions and can catalyse the attachment of nucleotides, but cannot carry out all replication, in addition it is joined by the beta subunit.

Beta subunits differ from other proteins in that they have their own enzymatic activity, which is unnecessary for DNA synthesis - ATPase activity. Beta-subunits bind ATP and hydrolyse it in very small amounts during synthesis. Further, an interesting phenomenon was discovered as a rearrangement of the structure of the betta subunit (indicated in blue everywhere in Figure 3.6) during ATP hydrolysis. The two subunits normally without ATP form a claw, contacting one end of each subunit. Once attached, this structure becomes sensitive to DNA binding. When DNA gets there, the beta subunits close into a ring due to ATP hydrolysis. In this way, the DNA enters the bark part of the enzyme (cor-enzyme), where the second strand is synthesised on it.

Cairns' research using autoradiography was able to prove that replication is a process in which parental chains are simultaneously unravelling and replicating. Cairns obtained radioactive DNA, covered it with a layer of photographic emulsion and left it like that for several weeks; during this time, radioactive thymidine left "traces" - grains of silver - in the photographic emulsion, creating a photographic imprint of the DNA molecule. These imprints show that the intact E. coli chromosome is a single giant ring structure 1 mm long. In radioactive DNA isolated from cells during replication, moreover, an additional loop was found. Cairns concluded that, the loop is formed by the formation of two radioactive daughter strands, each of which is complementary to the parent.

According to the data obtained by Cairns, both DNA chains replicate simultaneously, and various variants of his experiment showed that DNA synthesis starts at one strictly defined point, the so-called origin of replication, and further synthesis proceeds in two directions - each side is operated by a different enzymatic complex. Gradually increasing such a replicating part of DNA eventually turns into two separate rings (Fig. 4.7).

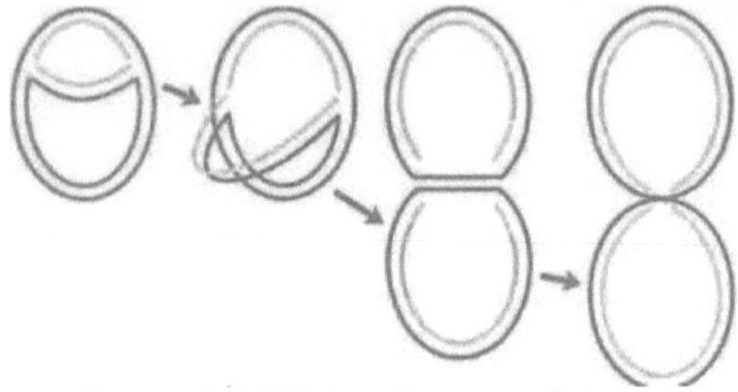

Figure 4.7. DNA replication of E. coli.

CHAPTER 5

5. Replication of prokaryotes and eukaryotes

Replication

The most important principle in DNA doubling is the principle of complementarity (geometric selection), the direction from the 5'-end to the 3'-end. The chains are antiparallel, and one keeps elongating, in which case polymerase III can make a continuous strand of millions of nucleotides. A the second is formed as individual Okazaki fragments. And the synthesis of each fragment begins with a seed - a special enzyme, which consists of RNA. Why from RNA? The beginning is always associated with low precision, there is no rigid fixation of the first nucleotide, and there is a probability of errors, so it turns out the following: it is necessary to make some primer by a system that can do it (and DNA polymerases can not), lengthen it due to the work of DNA polymerases precise, remove the primer and replace it with normal DNA nucleotides. This problem is solved by making the primer RNA with a lot of errors, i.e. it is easy to remove, not very good at joining, not very long, and completely different in nature. Apart from the fact that synthesis proceeds differently along the two strands, it proceeds along them simultaneously. This is done because DNA polymerases have a fairly complex structure with two active centres. Two centres - which elongate the molecules, and when synthesis begins, a different strand of DNA is attached to each of them. To one, which is synthesised continuously - the leading strand, and to the second - the lagging strand, which must unfold 180° to do so. When unfolding, a sufficiently long chain forms a loop, so that on one strand synthesis proceeds linearly and continuously, and on the other strand - with the formation of loops that change from time to time (Fig. 5.1).

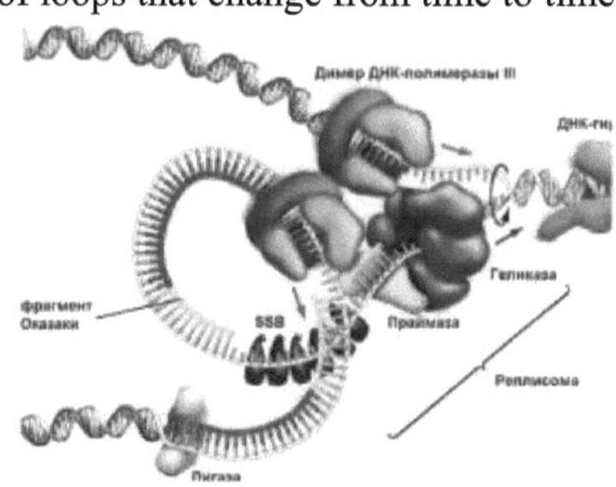

Figure 5.1. DNA loop during replication

Such synthesis is carried out according to the following scheme: polymerase synthesises a new strand, and two new strands are formed. The matrix DNA after polymerase unravels into two strands - a lagging and a leading strand.

Helicase binds to the polymerase and, as it unwinds, it immediately transfers the leading strand to the active centre. In this way the filament is synthesised continuously. The second filament bends and a loop is formed. Along this loop, the enzyme synthesises the filament in the opposite direction, pulling it through, but the result is a lagging filament that is completed. SSB proteins are displaced by the polymerase. In this way, an Okazaki fragment builds up, which will continue its synthesis until the enzyme bumps into the previous fragment. Single-strand breaks occur in the lagging chain. The polymerase reaches the primer, but cannot form a bond with it, and as a consequence, there is a break in the new chain. The old chain is continuous. Such a break must be closed. Here in different groups of organisms there are slight differences. DNA polymerase I is convenient for completing the synthesis. At the site of a single-stranded break, the inoculum ends in a 5'-end with one or more phosphate groups, and the newly synthesised site ends in a 3'-hydroxyl. They appear close to each other, DNA polymerase I, binds to the 3'-hydroxyl in this system, attaching nucleotides to it. The exonuclease activity, which is on a separate domain and can be removed by moving gradually along, detaches the inoculum and replaces it with DNA nucleotides.

DNA ligase, RNase, properties and functions

This process of single-strand break can last quite a long time, at least the inoculum is always eaten whole, then this process slows down, but it can continue for some time, and we have a pure DNA chain, but it is not cross-linked and still has a single-strand break. This single-stranded break is mended by another enzyme system that plays a very big role now in biotechnology and genetic engineering, which is DNA ligase. This enzyme mends single-stranded breaks. At the site of the cross-linking, there is most often a single phosphate. It has the bond that is needed, but it cannot join the hydroxyl because energy is needed. The energy source in the case of E. coli (E. coN) is a rather original nucleotide, nikotinamide-adenine-dinucleotide (NAD). Nucleotides are joined here not as in DNA and RNA, but by phosphate groups - a bond is formed between the two nucleotides.

two phosphates, and this bond is a macroergic bond. NAD (or in other cases ATP or GTP) interacts in the following way: the macroergic bond, breaking, serves to transfer one of the phosphates of the bond to the hydroxyl, as a result, instead of hydroxyl, we get AMP at the end, also attached through the phosphate, then the bond from AMP to the phosphate is transferred. So it doesn't matter which phosphate is bound, the energy will be about the same. The energy from the AMP attachment is spent on bonding to the phosphate standing next to it. Finally, the phosphate joins the two parts of the chain, the AMP comes off. The case where NAD is used is pretty unique, there are probably some other bacteria besides E. coli, but in all other cases ATP is usually used. If we talk about DNA ligases that are actually used in research practice and biotechnology, they are mostly viral DNA ligases. DNA ligases are typically phage T4, which uses ATP. Although the nucleotides are different, the reaction mechanism with ATP and NAD is the same, the only difference is the leaving group (nicotinamide phosphate or adenosine diphosphate). In this way, any gaps between Okazaki fragments that have formed are cross-linked. DNA polymerase removes the primer, builds up the resulting "gap", and the resulting single-stranded gap, which it leaves, due to the specificity of its enzymatic activities, is sutured by DNA ligase (Fig. 5.2).

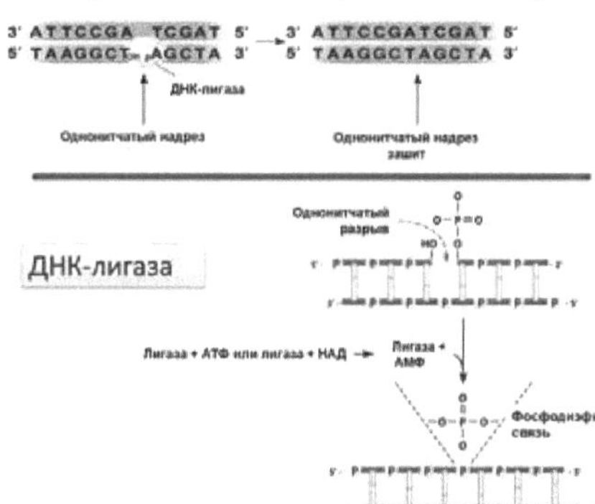

Figure 5.2. Principle of DNA ligase operation

The discovery of DNA ligase and the cross-linking reaction occurred before the mechanisms of DNA replication had been fully deciphered, but it was already clear that there would have to be cross-linking, i.e. there were Okazaki fragments. This enzyme was isolated after it was discovered, and this isolated enzyme turned out to be able to cross-link two DNA molecules if they have a single-stranded break, regardless of what it is, where it is and where this DNA came from, that is, the enzyme can do the reaction in a test tube. Genetic engineering started with the production of this enzyme. With these enzymes it is possible to cross-link various foreign DNA fragments, to make artificial constructs, and artificial constructs may be able to support DNA replication in some foreign cells. In order for the enzyme to cross-link, you have to have a single-stranded break. But in the early stages of genetic engineering, fragments were mostly produced by the activity of enzymes called class two restrictionases, which just produced such breaks if you joined the cut pieces together. This turned out to be very useful and very convenient for joining just such specific fragments obtained by restrictases. Next, another thing became clear. Ligase can cross-link not necessarily a system where there is one matrix strand, which is two separate fragments. It can cross-link any double-stranded DNA that fits together end-to-end, but it will take longer because the ends won't be fixed, and the enzyme sits on the end specially prepared for it, and you have to cross-link using a higher concentration of enzyme. When the enzyme became reasonably available, they started adding a lot more of it than they did for single-stranded breaks, and then they started cross-linking blunt ends.

E. coli DNA polymerase I is a unique case and other enzymes do not have 5'-exonuclease activity, so the problem arises how to remove the inoculum in these cases. It turns out that there is another enzyme in the cells to remove inoculum. It is called RNase-H (RNase-"ash"). Last time we said that there are endo- and exonucleases, DNAases, RNases, that is, nucleases with very different specificity. Nuclease-N is unique in its specificity. It's an RNase - it cleaves RNA. It acts mainly as an exo-nuclease, cleaving RNA from the end, although it can also cleave in the middle. And the interesting thing is that it doesn't hydrolyse free RNA. It doesn't hydrolyse double-stranded RNA, which viruses have. It only hydrolyses RNA if it forms a complementary double-helical structure with the DNA strand. It hydrolyses RNA in DNA-RNA hybrids. If there is a replication product in the form of Okazaki fragments with RNA inoculum, RNase-H sits on this place and starts to detach nucleotides and reaches the DNA, but since it is an RNase, it does not go further. The result is a gap, and DNA polymerase I of other organisms that don't have exo-nuclease activity or another DNA polymerase in the case of eukaryotes, which sits on that spot and starts adding nucleotides. It will get to the end of the gap, and it will leave a single-stranded gap where a DNA ligase will join in and sew up the gap. So, the ligase works the same way, but instead of E. coli DNA polymerase I, all other organisms have two enzymes: RNase-H, which acts in virtually all cells that replicate DNA, and DNA polymerase, which closes the gaps that form. There is another way to remove inoculum. There is a special form of DNA polymerase in eukaryotic cells that can, by sitting in a site with RNA, not hydrolyse the site, but push the site away because its energy of movement is quite strong. It pushes back by breaking the hydrogen bonds of that section of the chain. Nonspecific nucleases then break down the RNA.

DNA in prokaryotes is circular. The point from which replication goes both ways is located at a certain place in the membrane, and at this point a protein complex is assembled: proteins that recognise the replication point and that rearrange it in a certain way. proteins recognise certain sequences, bind to it and cause denaturation of the DNA and formation of a single-stranded site - this is easy to do when the DNA is intact because it is negatively superhelicalised. Two RNA polymerases are anchored at this point with a complex of auxiliary proteins, and there are thus two inputs: the DNA enters through the helicase into one active centre where replication takes place, and the same happens with the second part. It turns out that the two linked loops are at the same point, where the "replication machine" is located. It is worth realising that, contrary to the terminology, it is the polymerases that are fixed, and DNA is mobile. Due to the fact that replication goes both ways, each strand is both leading and lagging, both ends will eventually meet and can be cross-linked by DNA ligase.

Speaking about replication, it is worth to dwell on the peculiarities in eukaryotes. Prokaryotes normally have one circular molecule, where there is one starting point and an ending point at the opposite end. The replication rate can't be too fast. Because then errors will occur. Because of this, prokaryotes have delayed reproduction, and when bacteria reproduce rapidly, the replication time of the ring chromosome of the bacterium is longer than the cell cycle time. That is in E. coli, division occurs once every 20 minutes, but replication occurs once every 30 minutes. Replication, having started in one place, proceeds from the replication point, but there are several points in the cell. And when there is a part of DNA that has already undergone replication in these new molecules replication points are in the middle position from the work of this machine, these points are attached to other replication factories, and the second act of replication begins before the first one is completed. It turns out that the double-stranded ring is attached at this point to the replication factory and replication begins. After some time (the replication factory sits on the membrane), double-stranded unreplicated DNA enters it from both sides and two double-stranded DNAs come out from each side, then they close together. At the place where replication starts, a replication centre sits on them, so there is not one replication centre, but three. In the EM you can see first the appearance of an eye, the eye enlarges, the rest of it shrinks. New peepholes appear at the site of replication of new

DNA. There is a divergence of two rings that are half replicated. Thus, the speed of DNA replication in this case is measured not by the speed of a single centre, but by the speed of several simultaneously working centres: one works for the nearest cell division, and the others - for the subsequent ones (Fig. 5.3).

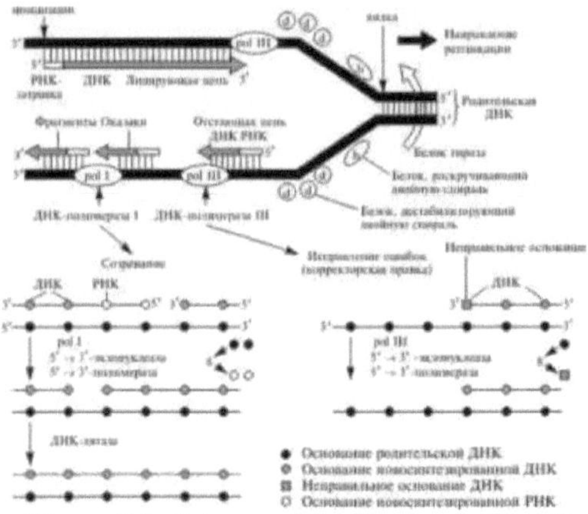

Figure 5.3. Scheme of replication in E. coli

Eukaryotes have a different characteristic. Their DNA molecule is nuclear - linear. Moreover, DNA is very large (tens of millions, hundreds and billions of nucleotide pairs). Eukaryotes have a mechanism of distribution of replication products - mitosis. Therefore, replication does not occur continuously, but in a certain S-period (1/10-1/20th of the cell cycle - about an hour).

Eukaryotic DNA polymerases replicate more precisely, but more slowly. And that's a problem. There are two ways to solve it: increase the enzyme activity, but this increases the number of errors. So the second way: occurrence of repeated replication sites before replication is complete. Eukaryotes have created multiple replication sites on each DNA molecule. Eukaryotes also have sites with complex proteins that carry out replication, replication factories, they also contain enzymes that unwind DNA, synthesise inoculum, carry out replication. Only such complexes are bound on the molecule not in one place, but in tens or hundreds of places. As a result, we get a molecule of a certain length with several sites of the same sequence so that the enzymes can recognise them, a replication factory is attached to each of them - replication begins. After some time from the point, some portion in either direction replicates, and an unreplicated site follows - and so on. Each site replicates at a small rate, but the total replication time turns out to be observable. These sites then meet. Instead of Okazaki fragments, large single-strand breaks are produced. There is no specific meeting site as in prokaryotes. Cross-linking also occurs by DNA ligase in the same way as in prokaryotes (Fig. 5.4).

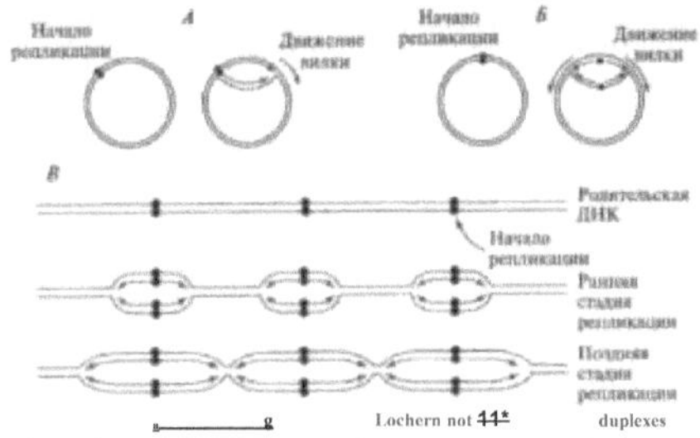

Fig. 5.4. Direction of replication: A - unidirectional replication in a circular DNA molecule, B - bidirectional replication in a circular DNA molecule, C - multiple bidirectional replication in a eukaryotic chromosome (po)l.irepl.11co)11)

The problem of terminal replication in eukaryotes. Telomeres, telomerase

It should be noted that at each site replication occurs according to the same scheme as in prokaryotes - in two different directions, one chain - leading continuous, and the second chain - lagging, in the form of Okazaki fragments with RNA primers. The RNA primers are removed either by polymerase, which displaces them by completing them, or by RNase-H, often both working together. There is another problem with the linearity of DNA. If a molecule is linear, it has the problem of end-to-end replication. This replication results in under-replicated single-stranded regions, 5'-ends, at the ends of the molecule. And this is the under-replicated Okazaki fragment. In the next act of replication, a complementary strand will be built up, but it will be shorter, because replication will not proceed at the broken end. As a result, with each new act of replication, the chain will be shortened gradually from each end. That's a loss of about 200 nucleotides. In the worst case, this threatens the loss of hereditary features (one gene is about 1 thousand nucleotides) and the final death of the cell. And here we should remember about the structure of the chromosome. In mitosis, it has telomeres. These are more densely packed chromatin sections at the ends. According to EM data, they're not just densely packed, they're packed differently than the central regions. It turns out that the ends of chromosomes in most eukaryotes are represented by repetitive sequences - and very specific ones at that. Unsymmetrical structure of a repeating element: in one chain A and C, in the other - G T. Such a sequence was determined in yeast. And at that time, genetic engineering of eukaryotes was just beginning to develop. And it's quite easy to introduce DNA into yeast by enzymatically breaking down their shell. They absorb that DNA quite easily after that. The relatively simple repeats that yeast had can even be synthesised chemically. They did the following thing: they combined such repeats with foreign bacterial DNA, introduced them into the yeast cell and added the yeast sequence needed for replication. It was introduced in the form of a bacterial ring plasmid, because there was not enough material, this ring molecule was multiplied in the bacteria, isolated from there and planted in eukaryotic yeast cells. It was expected that yeast would also have an end molecule, and specific proteins would sit on the telomeric end sites. But this did not happen: specific proteins sit down, but the molecule is not circular. That is, inside yeast, this multiple repeat was cut into two halves and shortened (there were three repeats on each side, and all the excess was removed). Next, they began to make six repeats, and the molecule was cut exactly at this point. Thus, a linear molecule appeared in yeast, and if everything there was still complemented with what was needed for replication, a normal small chromosome was assembled. This was used extensively. Because in that little chromosome you could insert a lot of DNA. foreign DNA, and it would replicate in yeast and exist as an extra chromosome. It turned out that yeast regulates the number of repeats at its ends. And if you sew a large number of repeats on the ends, and if you sew a large number into a ring, they cut by just cutting off the excess. If there's no extra, they just cut in half and you get two telomeres. If the yeast cuts off the extra, then that extra shows up. This extra is synthesised by the enzyme telomerase, it works on the principle. It contains an RNA molecule complementary to the telomere sequence. Behind the 3'-end of the DNA, which has ended, we have an additional RNA sequence, which serves as a matrix for telomerase to attach the following nucleotides to the 3'-end. Telomerase after it has synthesised what it can it pulls the RNA off, the RNA separates from the matrix and then the part that is at the beginning of the active centre joins the already synthesised one or two repeats. By shifting by these repeats. It gets again a 3'-ON group, which is free, and which has a matrix that can be attached. So by gradually shifting the RNA matrix, which is part of it and is always complementary because of the repeats, it will synthesise a certain number of these ends. There is a limitation. Yeast leaves only three double-stranded repeats on each side. The double-stranded repeats appear on their own. After telomerase has worked, a long single-stranded region emerges. A primer is made to it and DNA is synthesised that continues the strand. As a result of multiplication of telomeric sequences, a long section appears, on which the second strand is partially completed. Next, in yeast, the excess is cut off, leaving only this double-stranded portion of a certain length. Most other organisms do not have a process of cutting off the excess. They make a fairly large single-stranded piece, synthesise the second strand on it in part, and the specific single-stranded part is further folded. G- C and A-T are usually synthesised, but it can be different. Here we can take guanines, which form two pairs of hydrogen bonds in two ways. A these ways lead to the fact that one guanine can be joined to one guanine by two guanines on both sides. Thus, a guanyl tetrad is formed. It occurs when the single-stranded part, where there is no telomerase, starts to bend, and one guanine part joins the other. When there are many of them, they twist into a chain reinforced by such guanine tetrads. Certain proteins attach to these tetrads that stabilise and protect them. So this single-stranded end

is well protected, it is not broken by nucleases - they just don't get there. So during the existence of DNA before replication, the structure is stable due to these proteins. As soon as replication starts, the proteins fly off, telomerase comes into play, and it lengthens the strand by a certain section every cycle. And the telomeric

sequence, as it is seen, does not decrease, but can also increase (Fig. 5.5).

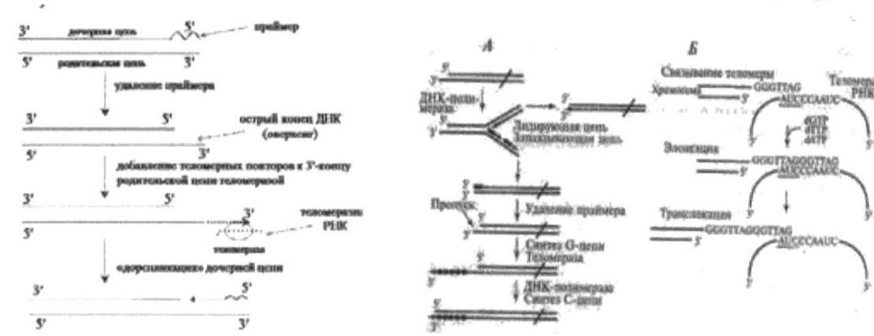

Figure 5.5. Telomerase

Telomerase-related cellular senescence

There is a phenomenon. Specialised animal cells, capable of dividing, divide on average 10 times and then die regardless of their state. It is assumed that this is due to the fact that telomerase does not work in these cells. And they really don't have telomerase. And cancer cells do have telomerase, so these cells are immortal, while fibroplasts, which we isolate from living tissue, can give ten generations of division, after which they die. More often than not, they die earlier, because you get fibroplasts in the body that have already divided several times. But in any case there is this phenomenon, and it was linked to telomerase, that the absence of telomerase makes it impossible to lengthen the ends, and ten divisions is about the size reduction that leads to the loss of vital areas. Hence the idea that if we could make telomerase work in cells, we would effectively make cells immortal, prevent the body from aging, and if we could learn to inhibit telomerase, we could learn to cure cancer. But it turned out that neither of these things is completely wrong. It is refuted by the following: a human can now live for 70 years, despite the fact that he has no telomerase, and a mouse lives a year and a half despite the fact that it has active telomerase in all cells. Attempts to force telomerase to form by inserting a specific gene in somatic cell culture did produce results when they crossed the 10 division barrier, but beyond that, once the barrier was crossed, they began to acquire changes characteristic of tumour cells. In fact, making telomerase appear in all cells and forcing it to overcome the lifetime of cell lines could lead to cancer.

Mobile elements. Replication by rolling end scheme

Speaking of telomere repair, one more thing is also worth mentioning. In insects and, apparently, other arthropods, the mechanism of end-sequence repair is different. It is based on so-called mobile elements. This is not so much the structure of DNA as it is the structure of the genome. In the organism, in almost any cell, there are certain sequences that are not particularly functional, but are maintained by reproduction and move around the genome. These have been called mobile elements. They were first discovered in bacteria. They were found to be useful transposons in bacteria, that is, as they moved from cell to cell, they carried the still important functions of antibiotic resistance. By these functions they were first discovered. There are many different transposons. The presence of mobile elements is present in all organisms, there are many of them, some of them originated from mutant viruses that entered the cell, embedded in the genome, but could not form a viral particle. We have about half of all DNA for this kind of sequences. Insects have elements that move quite regularly from the central part of the chromosome to the periphery. It turned out that in the same Drosophila, the end of chromosomes is associated with the presence of 4 different mobile sequences that move not randomly, but to certain points, and each of them finds this point in another sequence, and they seem to be embedded in each other in turn. When a certain piece due to under-replication disappears, it is compensated for by the fact that another gene - which is just such a mobile element - is built in a little further away. Although they have telomeres on their chromosomes.

A number of viruses and transmissible plasmids of prokaryotes have another type of replication - rolling ring replication. In this case, there is a special replication initiation site, which, unlike normal replication, is not unravelled, but where one of the strands is cut. The replication start point is ori. But in the case of plasmids (since they are transmissible) it is called ori T (from the word "1pp^psh1o11" - movement). A special protein encoded by this bacterium or virus recognises this site and sits there, breaking one strand, it does this in the same way that topoisomerase breaks a strand - it covalently attaches itself to the site rather than just breaking it. The DNA polymerase attaches to the other end of the break and starts to elongate the end, and the end connected to the protein is pushed back. Growing is the 3'end, pushing back is the 5'end. A second strand starts to form to the pulled back strand, and the enzyme can go full circle - the DNA that forms the second strand will be made anew, as it were. The polymerase will go on to the next round. A protein will come off, binding the newly synthesised section into a ring. In viruses, this will cut when the synthesis is over, but plasmids don't. Transmissible plasmids drag this resulting DNA end into a neighbouring cell. So there are proteins encoded in the genome of that plasmid that bind that cell with another cell that does not have such a plasmid, they form channels between them and pull this DNA through this channel. In this way, one cell infects the second cell with its DNA. It is clear that the plasmid is small, it will transfer there entirely and, as a rule, these transmissible plasmids do exactly that, and the latter usually carry the genes responsible for the transfer itself. If the transfer is interrupted for some reason, there will be no viable plasmid inside. This is due to an interesting phenomenon called the sexual process in bacteria, which has nothing to do with the sexual process in eukaryotes (Fig. 5.6).

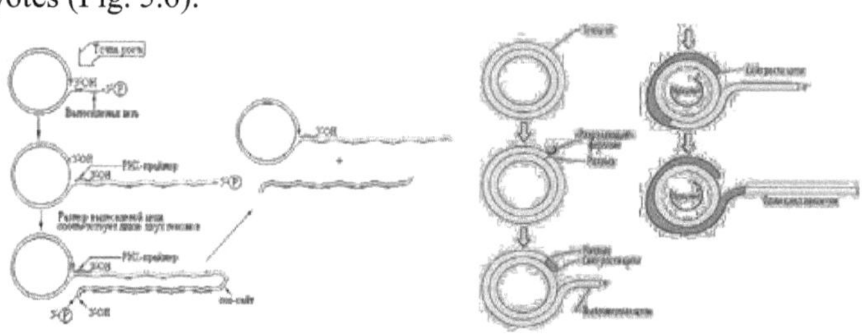

Figure 5.6. Rolling ring type replication

6. DNA recombination

In addition to DNA replication, there is also a process of recombination, which is the exchange of two DNA molecules at specific sites. The task is to create something new from an old set of certain alleles of genes. This process results in combinatorial variability. The most common type of recombination is the exchange of homologous sites (sites with similar sequences carrying the same genes, but differing in specific nucleotides and thus encoding different proteins). Such allelic forms are complementary in their main length (Fig. 6.1).

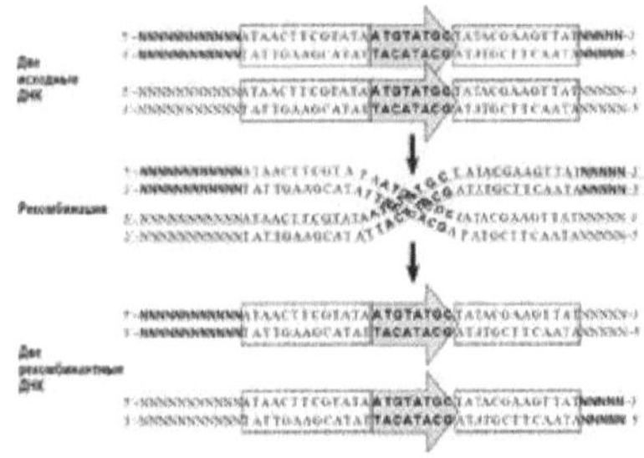

Рис. 6.1. Recombination

Escherichia coli Rec genes, transmissible plasmids

It turned out that certain mutations in organisms disrupt recombination: they worsen it or enhance it, more often they disrupt it. Such genes are primarily studied in Escherichia coli, and from the word recombination they received the designation Rec. They have a cruciform structure where there are two inputs and two outputs. The two complementary strands will at some point unravel and bind to another molecule of another pair. In hybrids, the sequences are not perfectly matched, there are mutations. Thus, it is possible in a DNA molecule to replace one allelic variant with another allelic variant. This structure is dynamic. If some diagonal ends are compressed and others are stretched, the point in the middle - the half chiasm - can move across the DNA. The process is non-energy-dependent because the connections between the parts are preserved. The movement takes place over fairly long distances - these will be the sites where the exchange takes place (Fig. 6.2).

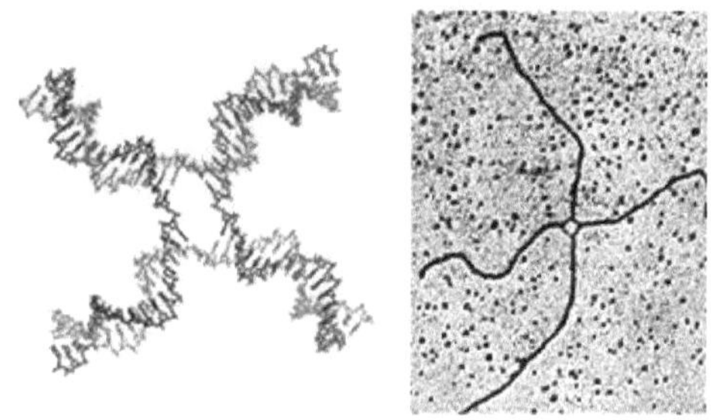

Figure 6.2. Homologous recombination. Right: schematically, left: electron microscope image

E.coli, for example, has a Rec gene system, there is the RecA gene, which encodes a small protein of the same name that has several activities. Firstly, it binds to single-stranded DNA. It is in the form of dimers and tetramers and coats the DNA in such a way that it presents a helix conformation - one strand is twisted into a corresponding double-helix sequence. This is possible due to the fact that in addition to the traditional Watson-Crick pairs can also be formed where the purine is perpendicular to the pyrimidine, thus being recognised by two pyrimidines on both sides. And the extra strand thus attaches to the double-stranded site and begins to displace one strand, forming hybrid DNA. This is an energy-dependent process: RecA spends ATP to carry out this process, so there is a possibility to replace the sequence in the old DNA strand with the corresponding new one, if there is no complete match, and get a new form of the gene (Fig. 6.3).

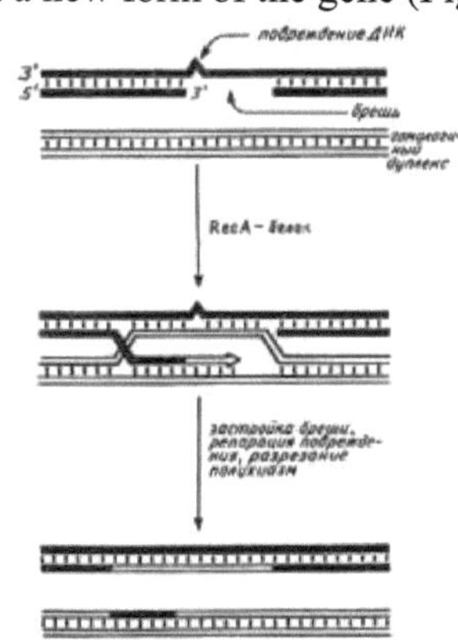

Figure 6.3. Mechanism of action of RecA

A complex of three proteins: RecBCD binds to the end of DNA - where it is broken. In prokaryotes, normally it is always circular. If there is a break, the complex sits on the break. If there is no break, it sits anywhere on the DNA

and waits for the DNA to melt. Then you get a fork, as in the case of DNA polymerase, and it moves along (energy-dependently) and the DNA behind it closes again. After a while the movement stops and the bubble that was formed will migrate along with the RecBCD complex. It does not migrate indefinitely: one of the subunits recognises a strictly specific DNA sequence. This sequence is called
"Hi" sites (ehi). It turned out that "Hi"-sites coincide with hot spots of recombination, i.e., sites where exchange often occurs. On the opposite side of RecBCD RecA starts to sit down, gradually closing this thread and preventing the formed bubble from closing. The second thread starts looking for another site. Once this site is found, a heteroduplex is formed: a double-helical DNA in which the strands have different origins, although similar sequence. This process of strand migration leads to the formation of a larger single-helix, which begins to interact with the strand that is displaced from the complex. A semichiasma emerges, which begins to migrate without RecBCD, and thus exchange occurs over quite large areas (Fig. 6.4).

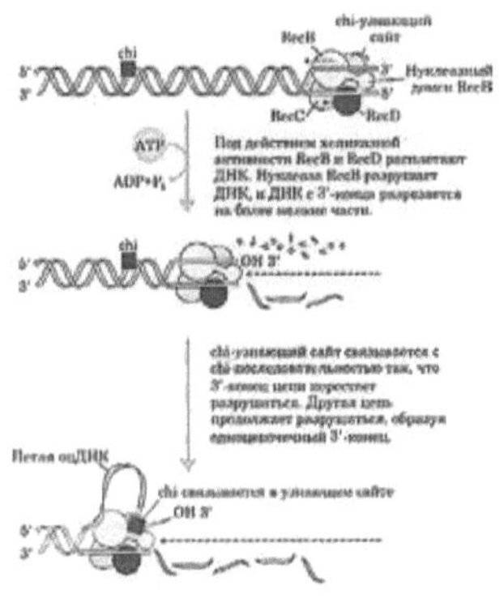

Figure 6.4. Mechanism of action of RecBCD

Bacteria *have* one chromosome, and if DNA replication hasn't occurred,
it's exchanged with transmissible plasmids. These are small
circular DNA molecules that encode several proteins,
to carry the DNA into the cell. It's transported
through replication by a rolling ring mechanism. These proteins are arranged so that they are further transferred to another cell, but this requires the cells to come close together. Among those proteins encoded in transmissible plasmids are proteins that encode so-called pilins - special proteins that form short outgrowths on the surface of bacterial cells (pili). A cell with a pili adheres to another cell, and between them the cell wall is lysed in a small area and the cytoplasm is joined. Some of the replacements that were in the

donor cell are transferred. This structure was first found in E. coli in the form of a large circular DNA. The protein it encodes, which in it triggers the replication of the rolling ring, can recognise not only the sequence in this plasmid, but just one point in the bacterial chromosome. Recognising it, it starts doing the same thing to the chromosome. A break occurs, a second strand is synthesised, and the separating strand goes off with the protein bound to it into a neighbouring cell. In this case, the plasmid is a donor not of its own DNA, but of chromosomal DNA. Chromosomal DNA starts to be transferred from a certain point, it is large, so it takes a long time to transfer, usually all the DNA is completely transferred in 90 minutes. That said, if there is no transfer process, the cells are just growing, they divide every 20-30 minutes. We were able to get a complete map of E. coli with all the proteins. Further sequencing of the complete genome confirmed that indeed there are indeed certain genes in it, grouped in certain groups, and their arrangement corresponds to that which has been determined genetically. This process in bacteria in terms of mechanism is not a sexual process - it is not the joining of two genomes. It is infection - one cell infects another cell with its DNA. DNA transfer can also capture the very plasmid that contains the set of proteins for transfer, and this plasmid is transferred faster, the protein works better with it than with the chromosome.

Strains with high recombination rates

There are strains with a high recombination rate - Hfr (high), and when the strain transfers its DNA to the recipient cell, it never transfers the plasmid itself virtually, because it turns out that the plasmid is sort of embedded in the chromosome of the cell. This plasmid was called the F-factor. It turned out that the plasmid and the chromosome have identical regions, thanks to which the transfer to another cell begins, but the breaks can lead to cross-linking of the molecules: a chromosome connected to the F-factor is formed. The chromosome is then transferred at the same frequency as the F-factor, and it is always transferred, meaning the chromosome is transferred quickly. It is also convenient that the recipient remains the recipient because the break occurs at a different site and the whole chromosome is transferred but not the F-factor, so something else can be transferred to the recipient. This is what is involved in recombination after DNA exchange in bacteria.

Recombination in eukaryotes. Homologous recombination

In eukaryotes, recombination is intense in meiosis, in the first prophase of meiosis, when chromosomes form pairs and homologous regions are brought closer together, crossed, and such exchanges take place between them. Speaking about such processes, it is worth noting that there are two variants

of what will result: there are two homologous DNA molecules. To start the recombination process, a break occurs, and the two strands are crossed, i.e. RecA-protein actually creates such a structure - a semichiasm is formed, which can be cross-linked, or migration of the cross-link along the molecule can begin, and then ligases cross-link the break. You get a structure in which one DNA molecule - one strand is whole, and the second one is represented by two fragments, and between them are fragments from another strand, and exactly the same on the second DNA molecule.
The molecules seek the position with the lowest energy, so they start to rotate, so two straight chains are bent and two crossed chains are straightened - this structure is called a Holiday structure. Next, this Holiday structure has to be cut. The process can follow two paths. It can be cut either horizontally, not along but between the chains, or vertically. The horizontal cut results in one whole fragment and a second with a break that contains a fragment complementary to the strand that was originally paired with it, the same fragment in the other section, and a small embedded fragment from the opposite strand. On the other side of the cut is a similar situation. Thus, the middle sections have been exchanged. Further, the gaps will be sutured, and molecules with non-identical in sequence insertions will be obtained (Fig. 6.5).

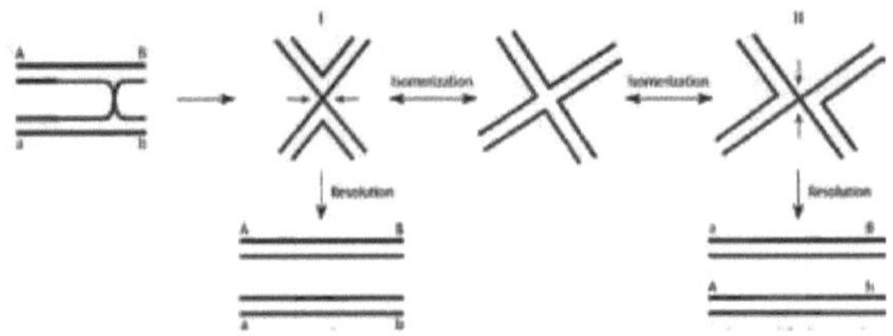

Figure 6.5. Holiday structures

If the cut goes in the vertical plane, the thread that was continuous - not torn at the beginning - will break. And with it in the complex will remain the thread that was going as a side thread - as a result, the thread where the cross-linking took place will be a continuous base, and the complementary thread will consist of two parts - from one and from another molecule. In this case, the end is exchanged. There may be several such exchanges, and as a result it may turn out that the end has not changed, but the middle has changed anyway. Almost any sections can be exchanged in this way. But the closer the sites are, the less probability of exchange between them.
In case of site swapping, a heteroduplex occurs in the middle and this cell is repaired. If the repair is random, then both copies remain evenly, as a result,

if there are few sequences (e.g. two), one of the copies may disappear, because each chain is complemented by a complementary one.
alien, thus producing identical molecules. This phenomenon is called gene conversion - the transformation of one form of genes into another.
In the case of exchange of ends, the area that undergoes repair is small and it is usually not very significant, so there is no loss of any form of genes there.
In eukaryotes, exchange is complicated by the fact that the DNA is in chromatin, but on the other hand simplified because there are proteins in chromatin that can help maintain such structures for quite a long time. Whereas in bacteria the migration and cutting of strands takes minutes, the prophase I of meiosis lasts for days, and all this time the chromosomes appear to be close together and held together. There are further peculiarities in this process in eukaryotes. For recombination to take place, it is necessary that the sequences fit well also in length. If it is small naked DNA of bacteria, then the same RecA protein moving the strand and applying it can find a site, where there is complementarity, and it will be fixed simply due to the formation of stronger bonds. But in eukaryotes it is much more complicated, because chromatin with a large number of proteins complicates these processes. And when preparation for division begins in the prophase of meiosis, the chromosome begins to condense, that is, the stacking is still compacted. How do chromosomes find complementary sites at such great distances with almost nucleotide accuracy? Presumably two things play a role here. One: there are highly repetitive sequences (satellite DNA). These are relatively short, not perfectly repeated sequences. H blocks that contain there 6-810 nucleotides can be repeated tens of thousands of times and can even form a blocky structure in some places. The types of such sequences and their mutual arrangement are different throughout. The second important point is that before entering meiosis, DNA replication occurs in the cells, as in the case of entering mitosis, but in myiotic cells DNA is not replicated to the end, there remains a quantity of unreplicated DNA scattered throughout its length. This is probably how chromosomes are fitted with identical sections to each other in meiosis.

Site-specific recombination

Another recombination mechanism: site-specific recombination. Initial studies were done on a bacterium of a virus, namely Bacteriophage lambda. This is an E. coli virus that is quite aggressive. Attacking a cell, it destroys it in a short time. Several dozen new particles are produced, and they infect the next cells. The structure of the virus is quite complex, its genome is average - 43 thousand pairs of nucleotides. In this genome, the ends are repetitive. This

structure is covered with protein, leaving the cell, binds to a new cell, destroys its shell, and DNA penetrates inside. And it turned out that if such a phage is grown for a long time on E. coli strains, that is, when the phage is grown, the entire E. coli is destroyed. It remains in a liquid medium, this liquid is used to infect a new culture of E. coli. If this is done without selection, then in parallel with the phage that multiplies, another group of cells is multiplied, which the phage does not destroy. Such resistant strains have attracted interest. If such a strain is grown for a long time, crossing on liquid medium without any selection, viruses begin to appear in this liquid, in which the resistant cells grow, although they were not introduced there. These viruses can be isolated and infect other unstable cells with them and multiply. Since only the cells were moved, it means that the virus was sitting in the cell, hiding. No virus was observed in the EM. If resistant cells are exposed to some life-threatening influences, they break down very quickly, and ready-made viruses are released from them. That is, there was no virus, it could not be observed by any genetic methods, but under unfavourable conditions the virus activated, killed the cell and went out in search of a new cell. Strains that spontaneously or under the influence of some factors are lysed and release viruses are called lysogenic strains. In these strains, the virus is in a latent form. The phenomenon of lysogeny occurs by itself when the environmental conditions are good for the cell: the cell is good, the virus is good, and some of these viruses pass into a latent form and multiply with the cell inside it. When the phenomenon of lysogeny occurs, phage particles appear, this phage can be isolated and its DNA can be looked at. It turns out that in some cases the phage grabs a piece of host DNA in addition to its DNA. If you infect the next cell with it, it can transfer a gene. It turns out that the phage only captures a certain gene - that spot,
in which the virus is incorporated into host DNA, forming a complex (Fig. 6.6).

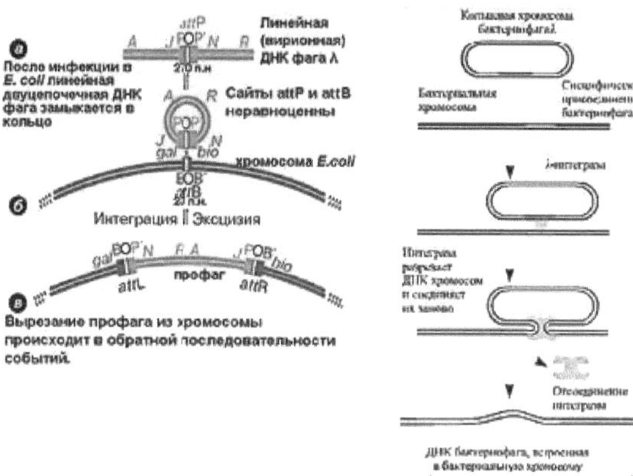

Figure 6.6. Site specific recombination of lambda phage

Viral and chromosomal DNA have similar sequences, which with the help of special proteins recognise each other and at this point there is a break in the phage and chromosomal DNA, which with proteins bind into a complex, after which the viral and host DNA are cross-linked. A loop is formed. The loop can unfold, if it is sterically fixed, topoisomerase comes into play, thus DNA is produced as part of the host DNA, covalently attached to it and is a continuous part of it. This sequence is also replicated along with the DNA. The sequence does not get bigger, the virus does not replicate in the cell, the cells containing the virus replicate, the viral particles are hidden in the cell. This is the strategy of the parasite: sit in the host and do no harm. When the cell becomes unhealthy, the opposite process conventionally occurs. The DNA gets twisted, bringing the viral ends closer together and forming a viral particle. It is worth noting that while the virus is in this state, it has genes working. Normal viral genes are designed to kill the cell and make lots of viruses. The gene that is working here is a gene that makes a protein that suppresses all other viral genes - a lambda repressor. As long as everything is OK in the cell, the lambda repressor keeps the DNA from being cut out and turning on proteins that will replicate it uncontrollably, keeps the envelope proteins from forming. That is, the DNA is conserved with the help of a repressor protein bound to the viral genes. In the case of certain influences, either this protein is destroyed or its synthesis is inhibited. Sometimes it is protein-related, that is, the virus can be activated by the RecBCD system because there are breaks in certain parts of the DNA or exposure to UV. In any case, phage exit is cell death.

Recombination takes place at specific points with a strictly defined sequence. This sequence must be repeated in the two DNAs involved in the process.

When the DNAs bind, the sequence by which they bind must not just be the same in the two molecules, but must also be mirrored within the molecules, that is, it is an inverted repeat (palindrome). There are subunits of the protein called recombinase. When recombinase binds to certain sequences, two identical sites in the DNA are created, to which two identical subunits bind. Once bound, they will do what topoisomerase II does: break the chain and attach its phosphate to its tyrosine, tyrosines attack the point between the binding sites and sew the corresponding 5'-end (the one with the phosphate) to itself. It will then appear that one of the strands is bound covalently to the tyrosine and the other has a free hydroxyl. The hydroxyl attacks the tyrosine bond to the other phosphate. The result is a structure similar to that of Holiday's: two twisted strands that can simply straighten out with a twist, making a single DNA molecule with a foreign piece embedded in it. Then the break occurs again, but other groupings will be attached to the tyrosine. The result of the break in the other plane will be that the molecules will sort of exchange ends. The ends will be non-homologous, they will be different sections of DNA. In the case of circular DNA, the DNA is cross-linked into a single ring (the case of lambda phage). In some viruses and plasmids, replication does not result in two molecules, they are linked together, one crossing over into the other, forming a double ring. To cut this ring, this mechanism is activated, working in the opposite direction, it will cut one DNA from the other.

Since the two molecules do not always diverge as a result of this process, catenanes, that is, two linked rings, are formed. When a catenon is formed, it can be resolved by site-specific recombination, or by topoisomerase. Here there must be repetitive DNA and their mutual arrangement leads to different results. One option is to have the sequences where recombination is taking place facing each other, with some piece of DNA in between. A central inversion of the genetic material occurs. Sometimes it turns out that this section contains the region where gene reading begins - the promoter. Depending on the direction of this section, the gene will be read either from one side or the other. This is how the process of gene switching can occur. This process is rare but vital. For example, this is the mechanism by which yeast change sex. Under unfavourable conditions, yeast form a zygote with a dense shell - this is what yeast of the second mating type are for. The genes responsible for mating are located on both sides of the promoter region.

In the case of direct repeats, in order for recombination to occur, the molecules must bend: not just form a loop, but cross over. As a result of the cross-linking, one repeat remains, half from one part, half from the other, and the other part closes into a ring, which also happens to contain one repeat. Such a process - the leaching out of a particular sequence - is called deletion. The example of the lambda phage is just such a process.

There are such systems that use not strictly complementary sequences for incorporation and can transfer the corresponding genes within such a ring to different places in the genome. This is how the so-called mobile elements function. They were first discovered in bacteria and were called transposons. Eukaryotes also have them and there are many more of them, but as a rule, they are not functional, and most often they are mutant viral sequences that have been inserted into the genome and cannot get out.

There are more complex mechanisms of transposition where replication works in addition to recombination. This is not just transposition to a new place, but doubling. That is, in the place where there is leaching, there is an exchange of two strands, and cleavage of one strand and completion of some site, which leads to doubling of mobile elements. About half of the human genome is made up of mobile elements. Proteins code for 5%, and they code for 50%. Such mobile elements in bacteria often carry antibiotic resistance genes. These genes multiply and transfer from one place to another, forming a structure that connects two transposons. This means that two antibiotic resistance genes for two different antibiotics are together. This gives them a selective advantage. Therefore, antibiotics should not be used frequently and uncontrollably.

Formation of antibody genes

There's a vital process - antibody gene formation. It starts with trypanosomes. They live in the blood and secrete toxic substances that can eventually kill a person. However, the human immune system works and produces antibodies against the surface protein of trypanosomes. They have developed a protein that assembles into a film that surrounds the entire cell. When antibodies are produced against this protein, the trypanosome rearranges the gene that encodes this protein. That is, trypanosomes have a particular section in the genome where the work of that gene is done, and that gene is assembled from pieces that are passed down from other parts. The person has to re-produce antibodies. Antibodies to these proteins are produced slowly and poorly. The same principle is used to form the parts of receptor proteins and antibody proteins that recognise specific antigens. The antibody protein is large and consists of several sites and polypeptide chains of two types: the variable site, which consists of three types of sequences: constant (all have the same sequence), variable and linker. In a normal cell, these genes do not work because they sit far away from the reading site, and they are not assembled.

In those cells that have become cells of the immune system and which are formed in the thymus during embryonic development, recombination processes begin. From a large block of variable sites, a certain one is selected and transferred to a constant one, and a linker (it usually consists of two blocks of some kind) is transferred there as well. Then such a gene starts to produce information in the form of matrix RNA as a result of the transcription process. That is, a certain molecule complementary to the site is obtained, which captures certain regions, but the connection of these fragments is not smooth. There are single-stranded joining sites between them. This is where the part is built up by a special DNA polymerase that only works in maturing cells of the lymphatic system. This polymerase builds a completely non-complementary system, and randomly builds 6-8 nucleotides into these binding sites. This site has many different variations and is recognised by antigens. And then that antigen manifests itself in different ways. It is a component of antibodies, but in addition, in other types of cells it is embedded in the membrane and serves as a receptor, that is, the immune response is given not by an antibody, but by a cell that has the same receptor, read from the same gene, but only it has a different back part, from a different molecule, and because of this it is embedded in the membrane. This is how a large number of variants are formed, in the thymus there can be about 10 billion variants formed. It goes differently in each cell, so each cell contains some type of antibody. What happens next is selection. Those cells that produce antibodies against their proteins in the embryonic period are destroyed. If the receptor has not bound anything, if there are no such proteins in the cell, then it will be needed to bind something foreign. And it persists. That's the thing about antibody formation in mammals. In birds it is a little different, but the principle is the same: there is also recombination of individual fragments. Olfactory receptors were formed in the same way in humans in the early period of development: that is, there is a huge number of types of olfactory receptors in humans, which are encoded due to recombination in different cells. It turns out that they are not originally present in the genome, and this situation is a bit more complicated overall. In any case, recombination produces variable products with certain specificity.

7. DNA repair

Causes of DNA damage. Ionising radiation

DNA, based on its functions, must be unaltered. The process of repairing damage in DNA is called repair. It can be subject to chemical and physical attacks. In terms of physical effects, high-energy particles are more likely to collide with the surrounding water than with DNA, so very often reactive oxygen species or active radicals like hydroxyl are generated, which can act on the DNA and then the chemical process begins. In other cases, such radiation can directly hit the DNA molecule and the energy-rich beta particles can cause the DNA to break. When bombarded by such a phosphate particle, the phosphodiester bond can break, then a single-stranded break is formed at that location. Alpha particles can cause double-strand breaks. However, at high radiation intensities, even more inactive particles can cause double-strand breaks.

DNA break repair. Oxidation and methylation

A single-stranded break can be eliminated by DNA ligase. When a single-stranded break has phosphate at one end (usually the 5'-end), the DNA ligase takes ATP/NAD and attaches the adenyl part to the free end and then, displacing it with phosphate at the other end, forms a bond. But only when a single gap is formed. It is also possible when the phosphate is not at the 5'-, but at the 3'-end. In this case, the ligase does not know how to activate the unphosphorylated 5'-end. So here two enzymes come into play, one of which is not very specific, in some cases there are even special forms of it for repair, it is called phosphatase and it detaches inorganic phosphate. This produces DNA that does not contain any phosphate at all at the ends of the single-stranded break. This is where a specific enzyme, polynucleotide kinase, comes into play. The enzyme sits on the 5'-end, uses ATP, releases ADP as a result of the reaction, and the 5'-end becomes phosphorylated. The ligase then sews up the gap. In the case of double-stranded breaks, direct repair is usually not possible. In this case, a combination of repair and recombination processes occurs. In the cell, as a rule, the same part of DNA exists in the cell, since replication is going on in the cell all the time, there is an exchange of strands. The strand with a free 3'-end will be built up by the second, as by the matrix into a longer one, then a single-stranded fragment, freed due to RecA protein, will be separated, embedded into the second broken section and will further serve for the origin of the second chain of the second half of the broken DNA. In this case, there must necessarily be a second similar sequence: the DNA must have already been replicated, if it is a bacterial cell, and in eukaryotes it can be the sequence of a homologous chromosome.

Cross-linking of double-stranded breaks is one of the most difficult cases of repair. Other cases of repair may be related to the fact that the gap may be not a single one, but a whole piece of the gap. In this case, Okazaki fragments are synthesised on the lagging fragment, as in replication, and then the primer is removed to form a single-stranded region, which is completed by DNA polymerases I and II.

In the case of more specific lesions, oxidation of adenine and cytosine often occurs. In this case, the non-ring amino groups of these bases are oxidised and replaced by a hydroxyl, isomerisation occurs, and a keto group is formed. And then in the case of adenine, after isomerisation in the six-membered ring, a keto-group is formed in place of the amino group, and in the sixth position, instead of the embedded nitrogen, an NH-group is formed. The groupings will be dissimilar in their ability to form hydrogen bonds: the amino group was the donor and the keto group is the acceptor of the hydrogen bond, and vice versa, in the sixth position the acceptor will be replaced by the donor. So the pair at this position will become noncomplementary. The resulting base is called hypoxanthine. It is a normal product of purine base exchange in cells, but it is of no use in DNA. Because if you look at the way it forms hydrogen bonds, you can see that it does it in the same way as guanine, and the complementary base would be

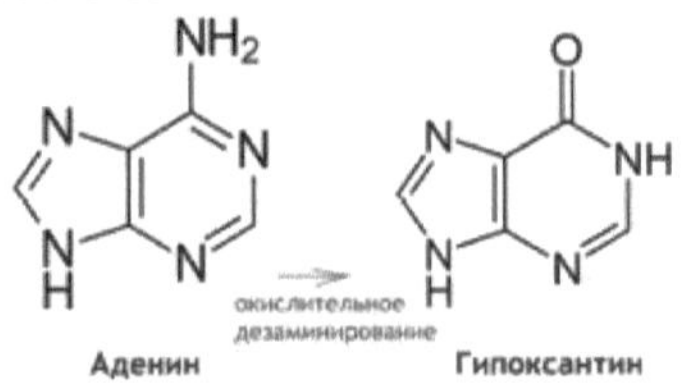

Adenine Hypoxanthine

cytosine. So, in the presence of this base in DNA, cytosine will stand against hypoxanthine - the replacement of one base for another and, as a consequence, the replacement of one amino acid for another in the protein, a point mutation is formed, associated with the replacement of one particular base and, as a consequence, one amino acid (Fig. 7.1).

base and, as a consequence, one amino acid (Fig. 7.1).

Figure 7.1. Oxidative deamination of adenine

If hypoxanthine can be distinguished from guanine, the situation with cytosine is more difficult. As a result of deamination, the amino group will also change to a keto group, and the NH group will replace the N- group, which represents uracil (Fig. 7.2).

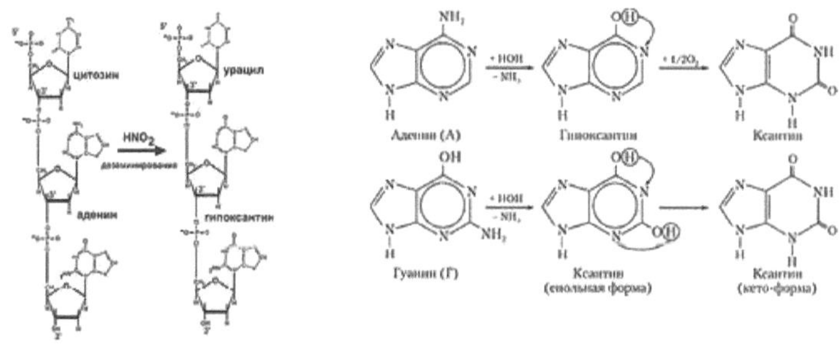

Figure 7.2. Oxidative deamination of bases

Then it becomes clear why uracil is replaced by thymine in DNA. If uracil were present in DNA as in RNA, such a mutation could not be distinguished from a normal mutation. Since DNA contains a methylated base, thymine, it can be distinguished from the one resulting from such oxidation. Uracil is quite easily oxidised by a number of influences: oxidative deamination agents such as nitrites, oxygen radicals formed from water by ionising radiation; active oxidants derived from cytosine itself when it absorbs UV light. Thus, uracil is formed in cells quite a lot. So how does DNA get rid of it? In cells, processes involving the addition of alkyl groups, most commonly methyl groups, are quite common. Methylation of the NH-group embedded in the six-membered ring results in the base not forming a hydrogen bond at this point. In some cases, a keto group may be methylated, such as in guanine. This reaction occurs spontaneously, but with the participation of active methyl group transporters in the normal metabolism of the cell. Then a methyl ester is formed instead of the keto group, which leads to a redistribution of double bonds. In this case, the possible binding is broken, DNA completely changes the structure at this place, and since the simple ether bond is quite stable, it is quite difficult to get rid of this methyl group. Such mutations are called 6-O-methylguanine - one of the most dangerous for cells (Fig. 7.3).

Figure 7.3. 6-O-methylguanine mutation

In addition to the various variants of oxidation of amino groups, oxidation of rings can lead to DNA breakage. As soon as a ring is no longer aromatic and flat, the bonding (DNA bonding is the interaction of flat aromatic rings) of DNA is broken. The strength of DNA decreases, and during replication, no nucleotide can be placed against such a ring. So, a random insertion can occur in this place, which will also be a mutation.

Causes of damage. Action of UV radiation, thymine dimers

The action of UV light is not insignificant. Cyclic systems of nitrogenous bases by their properties well absorb near UV (wavelength of about 260 nm). In some cases, bases excited by UV absorption, due to the increased reactivity of their electrons, form additional bonds, most often with water components or hydroxyl. Further, an oxidised position is formed at this position, which leads to ring breaking or the appearance of additional substituents outside the ring (for example, attachment to the nitrogen in the five-membered cycle in purines, since the nitrogen in the seventh position in them is very sensitive to some groups). The occurrence of such a positive charge leads to further rearrangements and, as a consequence, mutation. The most noticeable effect of UV on DNA is the formation of thymine dimers. This is the joining of not two nucleotides, but exactly two bases. This is the case when two thymine nucleotides are consecutive in the chain (Fig. 7.4).

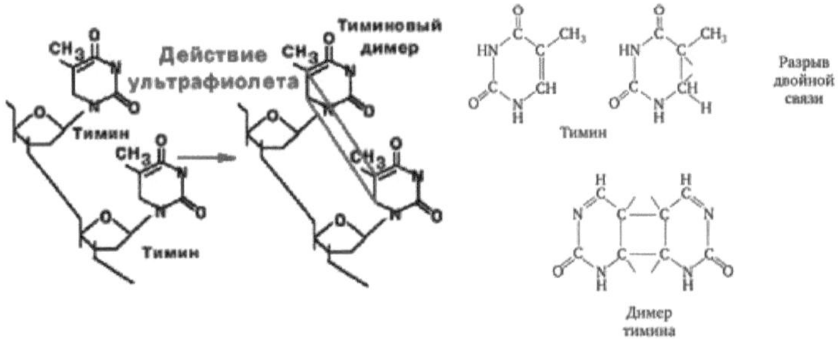

Figure 7.4. Thymine dimer. The formation of cyclobutane is indicated in red

Upon UV action, one double bond in the ring is broken and, as a consequence, it attacks the second neighbouring molecule, breaking its double bond. Such a cyclobutane dimer causes the two nucleotides to be strictly parallel instead of being rotated relative to each other by 36°. This leads to a violation of the overall geometry of the helix. In the case of replication of such DNA polymerase DNA will attach the first adenine, reaching the thymine dimer, and the second one cannot be attached due to the difference in angles, and replication ends there. Since UV is common, and many thymine dimers can be formed, it turns out that life is under the threat of such mutations all the time, and perhaps the presence of strong UV irradiation was the reason that until relatively recently life existed only in water. Water absorbs UV well, and at some depths it is no longer as strong. It is strongest on single-celled organisms and weakest on multicellular organisms with protective coverings. Under the influence of UV a person sunbathes, up to a certain point nothing happens. But still thymine dimers are formed in the cells, but the top skin cells are dead and the ones underneath are about to die, so it is relatively unimportant what happens to the top epithelial layer. The trouble starts when the cross-linking begins in that layer of cells that essentially forms the epithelium - the so-called basal layer of the epithelium, where cells are continuously dividing. It is clear that if such mutations have occurred there, the division will be disrupted. With some probability, the synthesis in this place still skips, sometimes cells deliberately make mistakes to pass such an unreadable part of DNA in order to survive. But when this happens, mutations begin in the epithelial cells, which very often leads to cancerous degeneration. This is why travelling south during peak season and intense sunbathing is not recommended. Hence the consequence: most people living in the south, in tropical countries, have a dark skin colour. Because their pigment cells are used to absorb excessive UV.

Photolyase, properties and mechanism. Pterin and its role

Speaking of thymine dimers, it is worth saying that this is a rather frequent and ancient phenomenon, and several repair (reparation) mechanisms have been developed against it. One of them is the direct repair mechanism. This is the repair of DNA damage that proceeds by strictly reversing the path of that disorder. If a thymine dimer is formed due to UV activation, it is also expanded using light energy, only in the visible region. The corresponding enzyme is called photolyase, and its effect was discovered rather accidentally. In some microorganisms, mutations turn out to be harmful for the microorganism and beneficial for humans. For example, there is a wild fungus penicillium that forms penicillin. If you take this fungus, as an English scientist did in his time, start to cultivate it and get an antibiotic from it, you will find an interesting thing: the fungus produces this antibiotic exactly as much as it needs, and it needs so much to suppress the growth of bacteria in the volume in which it grows, i.e. *1 active unit of antibiotic*. So penicillin produces penicillin with a concentration of 1 a.u/ml. For penicillin to have a real effect, 200,000 litres of culture medium is needed to treat one person. Cheyne and company in America learnt how to get penicillin mutations that were characterised by upregulation, where mutation is upregulation, so that the fungus does not make as much antibiotic as it needs, but almost uncontrollably. When they started getting mutations under UV, they found the following: if you take a suspension of yeast or bacteria, irradiate it with UV, sow it on solid medium to get colonies, when you disperse it in a dark place you get an order of magnitude more mutant colonies than in the light. It turned out that this is the effect of green light, but in many microorganisms this dependence is bimodal, that is, two pigments may be involved. It was assumed that it is flavin coenzymes absorb light. Then we simply started to look for enzymatic activity in favour of UV-irradiated DNA. And observed the melting point (thymine dimers lower it). In certain cases, it is possible to isolate an almost pure protein that has this activity. On the one hand, this protein has an affinity to DNA, but rather weak if the DNA is normal, and if the DNA has a thymine dimer, the binding constant increases by 6 orders of magnitude. That is, this protein will constantly attach to DNA and detach due to weak interactions until it binds to the site where the thymine dimer is located. Then longitudinal diffusion along the DNA occurs and the thymine dimer ends up at the active centre of the enzyme. This is the most favourable binding, which is what leads to the increase in the constant. Once the enzyme absorbs a quantum of light that activates its incoming flavin coenzyme, this coenzyme begins to interact with the thymine dimer, but it lacks energy. To increase its ability to activate carbon and expand bonds, a second chromophore, pterin (coenzyme), is used. The name pterin comes from the first place it was isolated - butterflies. It is due to pterins that butterflies acquire colour. It turned out that these compounds are vitamins, are present in all cells and are involved in the transfer of methyl groups. In some cases, they work as redox coenzymes, and it happens that they fulfil both functions. For example, when thymine is synthesised for DNA synthesis, the pterin transporter works as a donor of a one-carbon residue, and on the other hand, it restores a one-carbon residue to a methyl group. And so it turns out that pterin additionally activates flavin. It is necessary for two chromophores to bind one quantum of energy each, and then flavin begins to expand the bond, but it spends its energy to break one bond. A second bond is broken at the expense of energy, which he immediately receives at the expense of pterin by radiation-free energy transfer from pterin to flavin. As a result, the bonds of cyclobutane in the thymine dimer are broken, the thymines diverge to form a hydrogen bond, which leads to a complete reversal of the reaction (Fig. 7.5).

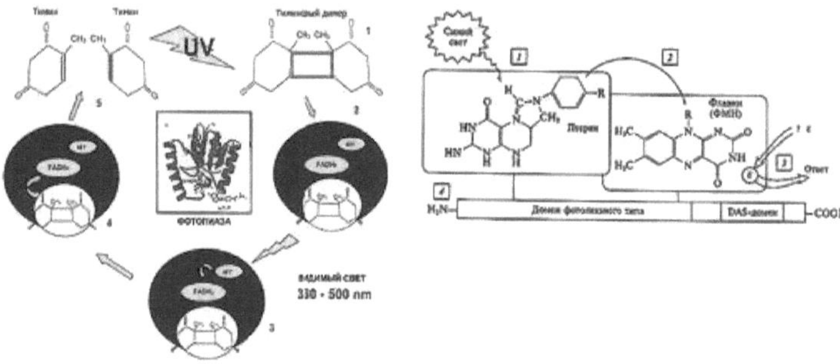

Figure 7.5. Mechanism of photolyase action during direct repair

There is another way to repair thymine dimers. The cells of eukaryotes have a special DNA polymerase that repairs thymine dimers. When a complex of two polymerases (alpha and delta), one goes on the leading strand and the other synthesises Okazaki fragments. When one of them bumps into a thymine dimer, it moves away, giving way to another polymerase that has a higher affinity for the thymine dimer. It's able to join two adenines. But it has a high affinity for thymine dimers, and once it has bound to them, it binds the two thymines together, and then the chain can continue to be replicated by the polymerase. Most often the thymine dimer is removed by a universal mechanism called excision repair.

In addition to thymine dimers, other unusual bases can also form, and there is a specific way of correcting incorrect bases when the whole DNA is basically correct.

Reparation of other disorders. AP site. Mismatch-, SOS-repair

In cases of other nitrogenous base disorders, there are two approaches: the first is specific to certain types of disorders. Certain types of disorders are either frequent (cytosine is oxidised to uracil) or too dangerous, and there is an individual mechanism of removal for them, just as for thymine dimers. Speaking of such mutations, we should focus on methylation of guanine, to 6-O-methyl-guanine. The mechanism of rearrangement is performed by a special protein O-6-methylguanine methyltransferase. The protein contains several active cysteine groups (SH-groups). The active centre of the enzyme grabs a methyl group, so that the enzyme is methylated by one SH-group and guanine is unmethylated. After this, the affinity of the enzyme for guanine is lost and the complex disintegrates.

Another interesting group of enzymes are DNA glycosylases. DNA with

oxidised cytosine (uracil), opposite it is guanine, an abnormal pair. Therefore, uracil must be removed to avoid stopping protein synthesis. Glycosylase breaks the glycosidic bond between the nitrogenous base and deoxyribose. A deoxyribose with a free reducing end, without a base, is formed. Such a site is called the A-pyrimidine site. With pyrimidines, the same process occurs even without the involvement of enzymes. Purine sites are formed as a result. Since the ways of elimination of such sites are similar, such sites have come to be called AP sites (A-purine and A-pyrimidine). Such AP sites appear under the action of other glycosylases as well. It is important to note that only one incorrect base is cleaved out, as the enzyme only recognises it. The most uracil is removed, then it is simply excreted with the urine, being methylated by nitrogen. At the time of insertion of the correct base, there is a difference in the mechanism of prokaryotes and eukaryotes. Eukaryotes have one mechanism and prokaryotes have two. The not-so-universal mechanism of prokaryotes is that the DNA insertase enzyme inserts the correct base from the nucleoside triphosphate at the A-purine or A-pyrimidine site. No such enzymes have been found in eukaryotes. There is a universal approach that prokaryotes and eukaryotes use. In the presence of structural inconsistencies in DNA, the site is excised and further built up. That is, the very fact of violation of some strand is recognised and this site is removed. This type of repair is called
excision (by cutting) (Fig. 7.6 a,b).

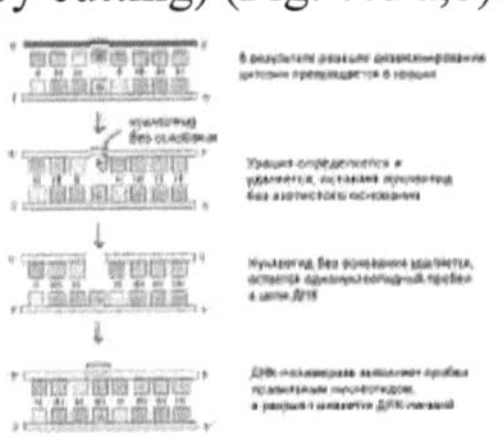

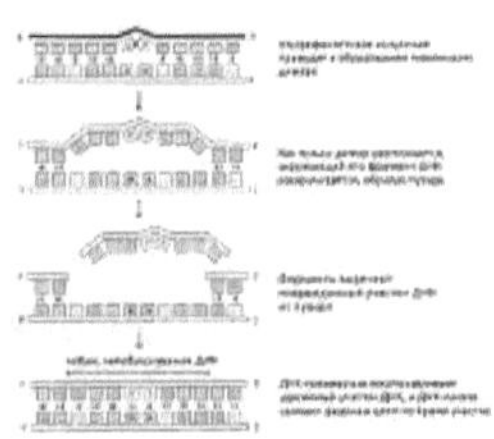

Figure 7.6. a. Excision repair of bases Figure 7.6.b. Excision repair of nucleotides

In this case, the uracil does not necessarily have to be deleted in the mutation, because the geometry of the double helix is disturbed at that location, and there are special proteins that control this disturbance. Once they detect it, they make a break in the "wrong" strand next to it. The result is DNA with a single-strand break and a mismatch. Next, another enzyme comes into play. It is worth remembering the difference between DNA polymerase of Escherichia coli, because this enzyme, unlike all the others, has an additional domain with 5'-3'-exonuclease activity. This enzyme sits on the single-stranded break and begins to gradually attach and remove nucleotides, passing the point of incomplementarity and destroying the mismatched site. In this way, the single-stranded break is transported along the molecule. Next, the DNA ligase sits on the single-stranded break and sews it up with ATP (NAD in E. coli). Since the others don't have the same activity of DNA polymerase I, another enzyme, exonuclease repair, comes into play. The exonuclease sits on the wrong site and starts to remove nucleotides from there, as a result a gap is formed - the DNA site will be deprived of the second strand too. DNA polymerase I or II continues the process, building up the gap, when it reaches a single-strand break, it stops and DNA ligase completes the process.

DNA polymerase works accurately, but it makes a certain number of mistakes. The number of errors in the replication process decreases dramatically after a while because cells have a special mechanism called Mismatch repair (repair of incorrectly paired bases). In a replication error, it is impossible to know which of the strands contains the correct base and which contains the incorrect base. The strands have different origins: one is an old matrix strand that was synthesised in a previous replication cycle, while the other has just been synthesised. Importantly, there are certain base modifications in DNA. In particular, in E.coP. If a quadruplet, which includes all the nitrogenous bases, is found in DNA, a methyl group is attached to the nitrogen of adenine, and N-6-methyladenine is obtained. The second strand has the same sequence with methylated adenine. After replication, the strands will separate, and a complementary chain will be built to each, in which adenine is unmethylated, because the newly synthesised strand has not yet had time to be methylated. And at this point in the cell, special proteins recognise this sequence and bind to it. Since methylation has occurred on only one strand, it is asymmetrical, and some protein subunits bind to one strand and others to the other.

So there is a structure that distinguishes the two strands, there are several such sites throughout the molecule. And this complex of proteins connects to another protein that binds them together, and also binds to the site between them forming a loop on the DNA, leaving the unmethylated strand on the outside. If there are changes in it, they are repaired (Figure 7.7).

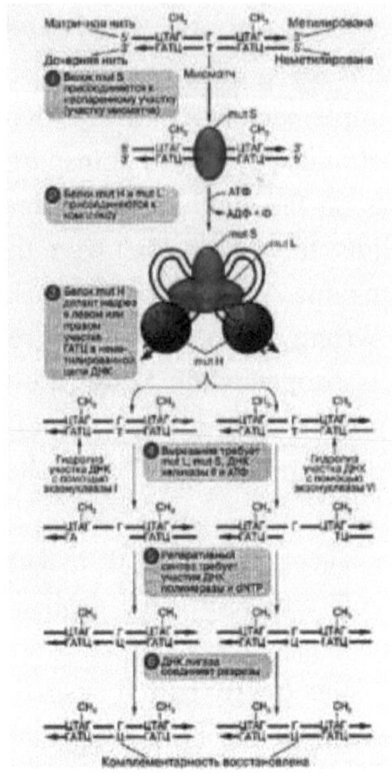

Figure 7.7. Mismatch repair

Thus, the presence of methylation as a signal on the old strand gives time for such errors to be corrected.

All of the above methods involve DNA degradation. Under conditions of high mutagenesis rate, SOS repair starts in cells. In the example of E. coli, the RecA protein is a nodal protein. It recognises single-stranded DNA, forms a complex with it, protecting it, and then begins to search for its complementary among double-stranded DNA and integrates the single-stranded section into the double-stranded molecule, displacing one of the strands, which forms a loop. The complementary DNA is synthesised on it, thus forming a double helix. The helix will have build-up gaps. It is important that during such processes, pieces of DNA fall out at closely spaced single-stranded sites; they do not linger because there are no long complementary sites. Such single-stranded gaps become dangerous for damage by nucleases. To prevent them from being damaged, there are special proteins called SSBs - they stabilise the single-stranded gaps that occur in DNA, then give way to DNA polymerases that build up the gaps. SOS repair requires genes, in particular the RecA protein, to be switched on at full capacity. RecA triggers all the processes, and normally it is not very abundant, its gene is in a partially repressed state. This is done by a specific protein that binds to it at the start of this gene and prevents it from starting

RNA synthesis. This is the LexA protein, a product of the gene of the same name. When RecA is free, it is inert with respect to other proteins, but if it binds to single-stranded DNA, its conformation changes slightly, it has protease activity that cleaves the LexA protein. LexA is a repressor not only of RecA but also of other genes - SOS repair genes, among them SSB proteins and some other repressor genes. There is one interesting group of genes: UMU (unusual mutation). They were discovered not in connection with SOS-repair, but because mutations in these genes lead to an increase in the number of mutations. The UMD and UMC genes, forming a complex, attack DNA polymerase, which is promiscuous. If it goes along the DNA and bumps into the wrong base, to which there is no complementary base, it inserts any. This allows the integrity of the chromosome to be preserved with sufficiently severe mutagenesis (Figure 7.8).

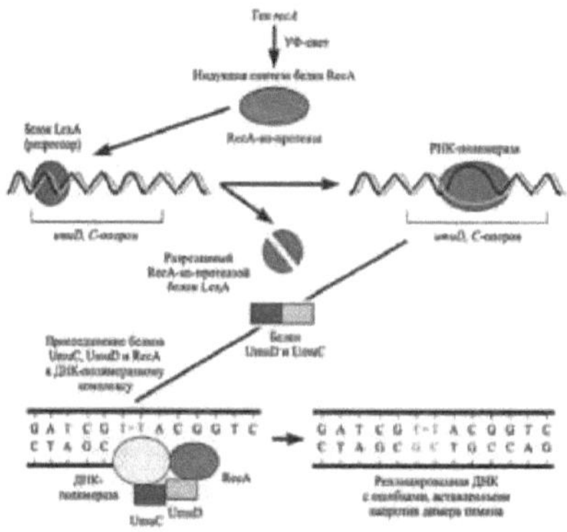

Figure 7.8. SOS-repair

8. Restriction systems. Realisation Genetic information

Of great practical importance is the phenomenon discovered back in the mid-60s by the Swiss virologist Werner Arber, who investigated the infection of various strains of Escherichia coli with a certain type of bacteriophage.

Host restriction. Arber's experiments

E. coli has a large number of natural variations. Certain such natural groups of strains have certain common features of structure (presence of some specific genes in these strains, a certain order of them, which may be different in others). They manifest themselves phenotypically not only at the level of genome research, but also in cases when it is associated with the use of certain substances.

There are a total of three lineages from which different strains of Escherichia coli for genetic research are derived. They are named after the strains that were first discovered in this line. Thus, there is a line called line "b" (E. coli B), there is a line named after the first strain discovered - it is line K-12 (E. coH K- 12) and there is line "C" (E. coH C).

By studying these three lines, Arber found that if you take one of these strains, infect it with a virus, the virus will penetrate the cell quite quickly and destroy it, while its resources will be used to form new viral particles (about 20-40 different viral particles will be formed), which will infect the next cells and so on (until only viral particles are left).

The researcher took a virus grown on one strain and infected cells of another strain with it, and it turned out that the development of viral infection was very delayed. Further this process was reproduced on a solid medium: a solid agar-agar medium with nutrient broth was taken, on which bacteria were sown solidly (the so-called bacterial lawn) and then a liquid suspension of the virus was placed there (viral particles are quite rarely located there). As a result, the solid lawn began to show transparent zones (so-called plaques) where the virus had eaten the bacteria. Gradually, the diameter of these plaques began to expand, and over a moderate period of time, one can observe the appearance of a large number of transparent viral colonies. These are those viral particles that are able to infect the cell and further develop. These colonies corresponded to the approximate number of viral particles of the suspension. At first nothing is observed, but then after some time, viral plaques appear, which are about two orders of magnitude less than a normal infection. From this study it was concluded that unlike the liquid medium, a small number of plaques develop on the solid medium.

Arber made this study with different strains and the result is that if we infect viruses grown on the same strains ("B", "K-12" and "C") and take the infection of the same strain as one hundred per cent, in the case of the "B" strain we infect the "K-12" strain, we get about one per cent; if we infect "K-12" with strain "B", we get about two per cent; it is noteworthy that strain "C" infects all forms of the virus with one hundred per cent efficiency and strain "C" infects strains "B" and "K-12" with rather low efficiency (less than one per cent) (Fig. 8.1)

X^fag E.coif[4]	B	K 12	C
B	100	~1%	>1%
K 12	~2%	100	>1%
C	100	100	100

Fig. 8.1 Infection efficiency of E. Coli strains

If you take the viruses from the formed plaques and infect the next bacterial lawn of the same strain (~1% from the table and infect strain "B" with it), the infections will go one hundred per cent. In fact, by multiplying once on this strain, the virus has acquired its properties for itself. It became one hundred per cent infectious. This phenomenon was called host restriction or host restriction.

It was only in the middle of the 70s that researchers came back to this discovery. They decided to determine what could be different in viruses grown on different lines of bacteria. It was determined that they had different amounts of modified bases, that is, certain nitrogenous bases became methylated at certain positions (the methyl group was derived from the amino group of adenine or the cytosine ring). These minor bases were present in small amounts, so initially it was not possible to detect them in any specific way, but it was possible to establish that methylation occurs along certain DNA sequences. Thus, the strains "K-12" and "B" were found to have different sequences.

Proteins involved in restriction. The first type of restrictases

Further research was aimed at finding the protein that was involved in this process. Initially, this was done using genetic methods, i.e. they started to get mutations that did not show such a reaction and it turned out that in order for this system to work, three genes were needed. Further it was found out that these are three proteins, which have different activities.

The first protein is the so-called recognition protein, which recognises a particular nucleotide sequence and binds to it. The other two are associated with this protein and have enzymatic activities towards DNA. One of these enzymes turned out to be a methylase, meaning it methylates DNA at or near these recognised sequences. The second enzyme turned out to be a nuclease, which only cleaves unmethylated DNA next to that sequence (methylated DNA is not cleaved).

Thus, the following scheme works in the cell: DNA is methylated at all sites having a certain sequence. That is, there is the original parental DNA and it is methylated (Fig. 8.2 (1)), then there is DNA replication, where the old strands are methylated and the new strands are unmethylated (Fig. 8.2 (2)). The enzyme, with its recognition subunit (denoted by the letter R), binds to these sequences, to which two subunits are associated, one a methylase (M) and the other a nuclease (N). If the recognising subunit has bound to the methylated one strand of DNA, the methylase comes into play and attaches the methyl group in the opposite direction. It turns out that the new strand is similarly methylated (Figure 8.2 (3)), after which the enzyme complex falls away from the DNA; if the strand is methylated at both positions, it will not bind. In normal replication, this pattern will always take place.

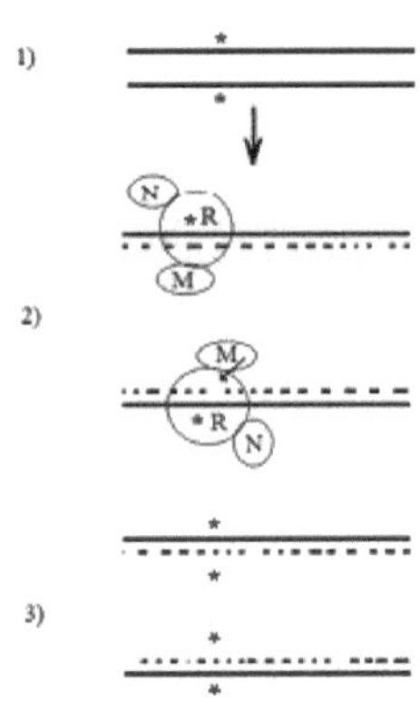

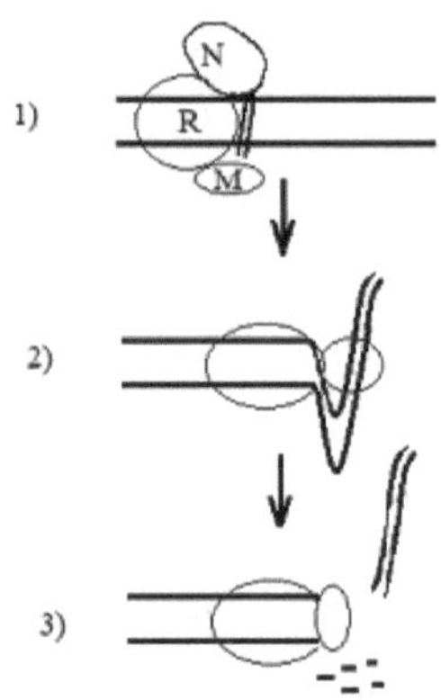

Figure 8.2 DNA methylation sequence Figure 8.3 DNA cleavage by a nuclease

If the virus is from another line of E. coli, it has no recognition subunit, and the sequences are unmethylated, then in this case, even before replication starts, the recognition subunit will sit where there is no methylation, and the methylation subunit will not be active, while the nuclease unit will be active and will start to destroy DNA (Fig. 8.3. (1)). Moreover, the degradation occurs in a very interesting way: the recognising subunit does not change its position, but the nuclease subunit starts to pull the DNA through itself, and a rather long loop is formed, which gradually increases. After some time, multiple bonds to neighbouring nucleotides begin to break (Fig. 8.3 (2)). The

result is that some DNA is left behind, then a complex of subunits, followed by a space with individual nucleotides between them (which are not paired and not held together) and then some DNA (Fig. 8.3 (3)).
Thus, it turns out that a large piece of DNA has been lost, which means that these sections cannot be restored in any way, and since there are many of them, the entire viral chromosome breaks up into small fragments. This is how the cell's defence mechanism works against the development of foreign viruses. But since this system does not work one hundred per cent, the virus tends to break through this system. Further, if methylation has passed on one strand, then there will be no more cleavage and it will be methylated on the second strand. If the defence system does not work, the virus is no longer identified as an alien and can multiply unrestrictedly. Since the phenomenon has been called host restriction, the corresponding enzymes have been named restrictases.
It was soon discovered that such defence systems exist in different types in different bacterial species. It turned out that restrictionases are of three different types (Figures 8.2 and 8.3 - mechanism of action of the first type of restrictionases). The first variant recognises some asymmetric sequence in the DNA and, if it is methylated, domethylates the second strand. If it is simply methylated on both strands, it is not retained, but if it is methylated, it cleaves it, and it cleaves it over a large length. This is a rigid system for destroying foreign DNA.

The second type of restriction systems. Use in genetic engineering

In most cases, a second class system works, where a relatively small sequence is recognised where the breaks occur. The enzyme closes a specific part of the DNA and interacts with it.
Suppose there is some section of DNA where the sequence ^{5} is -GAATTC-^{3}, in the second strand it will be complementary but oppositely directed - CTTAAG ^{5}. The sequence will be recognised on either side in the same way. After binding, if the DNA is not modified in this sequence, then at a certain point the subunit makes a break and since they are symmetrical, the break is obtained in both strands, but if the break occurs not in the centre of the recognised sequence, then a certain shift is obtained and a certain short part is formed, in which, due to the retained base pairs, as if this sequence is held in place. But four nucleotides is a very weak bond, so the retention is apparent and at normal physiological temperatures due to thermal movement it is quite often destroyed and as a result each of such ends will be autonomous
$^{5,---}$GAATT-^{3,,}
^{3}'-C5' where everything is complementary and in the end a kind of single-

stranded tail will hang. On the other thread there will be exactly the same picture, only in the opposite direction:

5C---$^{3'}$

^{3}TTAAG -5, that is, tips that are complementary to each other occur

No matter what the sequences at other sites - the ends will always be the same and complementary to themselves, which will stick to other ends from other breakpoints regardless of the sequence. Such structures have been called sticky ends by . The discovery of this mechanism and the use of such enzymes became the basis for the development of genetic engineering. This technique of DNA cleavage by an enzyme from the second class of restriction enzymes made it possible to join the resulting fragments along the sticky ends and obtain some new structures. This technique, called the recombinant DNA technique, is one of the main techniques for joining heterogeneous DNA.

Not all second class restrictionases make breaks somewhere aside, some tear exactly in the middle and then so-called blunt ends are obtained, i.e. two strands end at the same place, they cannot stick together, but due to the fact that in this case a phosphate group remains at the 5' (five bar) position of deoxyribose, they serve as a substrate for DNA ligase. When the sticky ends are joined, a single-stranded break occurs, which is also a convenient site for DNA ligase binding. That is, the advent of restrictionases and DNA ligases as analytical tools has led to the fact that it has become possible to join different DNA fragments. In some cases there may be several types of restrictases in one cell, in other cases we observe that bacteria do not have such restrictases (of the second class), they work at the expense of restrictases of the first class, but the joining of sticky ends is usually random and in order not to join all in a row do the following: let's say that DNA is taken from two different sources, therefore, one DNA is processed by restrictase and it remains as it was, and on the second DNA after processing by restrictase the end phosphates are removed, which means that such molecules cannot join each other and ligase does not work in this system. In the case when DNA with 5' (five-stranded) end phosphorylated is taken, such molecule can join with it and as a result only fragments from two different sources are joined. Considering that usually such constructs are made not from the total DNA of a bacterium or virus, but from a specially designed construct. Often this construction can be taken in even larger quantity, so that it could be joined with DNA of another source and at the same time have only one break point, then it turns out that in one DNA of a bacterium or virus different fragments from DNA of another organism are embedded in a certain place. This is how so-called genetic libraries are created. In order for such a construction to produce something, it must be introduced into the cell. For bacterial cells we considered the corresponding mechanisms (mechanism of genetic transformation), which at one time served for the discovery of the genetic role of DNA and this mechanism slightly improved is still used today. Bacteria are treated in a certain way to improve their permeability (the points where calcium ions are bound which sorb DNA well). The DNA to be injected is diluted in a suitable buffer, often with divalent cations added. Cell preparation conditions where the cell has a substitution on the surface of the ions and disruption of the charge in the envelope causes the uptake of DNA to become more efficient. A lot of DNA is added and the cells are further dispersed where genetics will already be at work. In this way, colonies are obtained, where different methods are used to select them depending on whether there is an insertion or not. It is important that these restriction enzymes make it possible to join heterogeneous DNA fragments by creating sticky ends. The sequence that is recognised by the second class of restrictases has typically four, five or six nucleotide pairs. Often, restrictionases that recognise different but similar sequences create identical sticky ends.

The third type of restrictases

The third class of restrictases combines some of the properties of the first class and some of the properties of the second class. They recognise a relatively small asymmetric sequence, just like the first class. They make a break at a strictly defined location according to the principle of a certain distance from the recognised sequence. If the first class makes a long loop and then destroys a very large fragment, here there is a break of only two strands with a small shift (usually by 2-4 nucleotides) as a rule not in the sequence in which it is recognised, but with some shift from it. That is, if there is a recognisable sequence, the enzyme sits on it and makes a break there. Single-stranded tails are produced again, but since the sequence is asymmetrical, they are always produced strictly at one end of the sequence and in a strictly defined location. The sites formed will be unique for each break point. The resulting ends are different for each point, which makes sense in some cases. So, for example, when you need to take a large DNA and isolate a small fragment from it, you can take and treat it with class 3 restrictionases, then treat one of these fragments with a class 2 restrictionase and remove the piece from there, and then already reassemble the molecule, because the sticky ends at each break will be different and when everything is mixed together again they will be reassembled in the same order.

Restrictase-like systems

It should be noted that there are some cases when cells have enzymes that recognise some of their sequences, in which gaps are introduced, which is used in recombination processes. That is, there are hot spots of recombination where breaks are introduced on a principle similar to that of restrictionases. A similar system is used by some viruses to infiltrate the host DNA. In all these cases, there cannot be many such points in the genome, so this kind of system recognises, as a rule, rather long sequences. The specific nucleases responsible for recombination points tend to recognise more than eight nucleotides in the sequence, which are used to allow some kind of recombination process to occur at one strictly defined location in the genome. Sometimes these kinds of breaks are used as a regulatory process. It was mentioned earlier that turning a particular site, for example in yeast, results in a change in the type of mating, that is, the site containing the promoter region of the gene's start site is orientated either one way or the other as a result. The effects are exactly the same for named nucleases, especially in viruses, which can switch some of their genes depending on the host strain. Such enzymes have been called homing nucleases (1yutschd nucleases).

The task of the cell is that it needs to make the right proteins in a strictly defined quantity and for this purpose a certain sequence of amino acid combinations is created. This sequence of amino acids is encoded as a sequence of nucleotides in DNA. The following contradiction was immediately recognised: eukaryotic cells DNA is in the nucleus and protein synthesis takes place in the cytoplasm. Initially it was expected that where there is DNA, synthesis would take place in chromosomes, but it coincided with the time of spread of electron microscopy method, and they began to make so-called electron radioautography. Some object was taken, a radioactive marker was injected into it, which was either bound to a certain place or localised only in some parts for a short time. In this case we are interested in the situation when radioactive amino acids with some speed were included into a cell and if we pick up a relatively short time and fix such cells, then we can observe that the tag started to be included into proteins, but practically whole radioactive proteins will not be formed yet. If in such a state the cell is divided into separate components, i.e. the cell membrane is gently lysed and the organelles that are there are further divided and after getting rid of the nuclei, it is possible to see how the radioactivity is distributed. Radioactivity will be mainly in cytoplasm, but proteins can be precipitated with acid or bound in some other way and this protein fraction can be measured.

The protein was expected to be synthesised in the nucleus, where the DNA is located, but it turned out that all the incorporation takes place in the cytoplasm. Then the method of electronic radioautography was used, which consisted in the fact that the cell was not destroyed after inclusion of the marker, but fixed, thin slices were made from it for electron microscopy, placed on a special substrate and further in the dark poured photoemulsion. Over the place where there were radioactive particles under the action of particles (carbon C14) a small amount of film was illuminated. And before this was done, the soluble label in the form of amino acids was washed with acidic solution. After incubation took place, the emulsion was subjected to the usual

and the silver grains above where the radioactivity had accumulated were produced. They were the same scale, и over the nucleus of the markings virtually hardlywas observed, It all ended up in the cytoplasm, where it was associated with certain granules that were visible in the electron microscope. These rather standard granules were found in all the cells studied, and hence the idea arose that these very structures of these granules were local protein synthesis.

RNA and DNA in systematics. Features of the Archaea group

A very reasonable question arose, if DNA carries information and sits in the nucleus, where does the information for synthesis in the cytoplasm in these granules come from? Then the concept of an intermediary or messenger that carries information from DNA to the site of protein synthesis was put forward. Such an intermediary is RNA.

Indeed, those particles in which the inclusion of radioactive amino acids was seen are largely composed of RNA (in the case of bacterial cells by 2/3, in the case of eukaryotic cells about 50%) There was this concept that there is a nucleus and in it there is DNA, which transmits (rewrites) information to RNA. RNA in its turn goes to cell to synthesise protein and, coming to the cell, it forms these protein synthesising particles, and each such particle synthesises its own particular protein. This concept existed for quite a long time, it was believed that the bulk of RNA that we see in the cytoplasm is a matrix for protein synthesis, but later it turned out that this is not quite true.

In the cytoplasm there were particles containing a lot of RNA, in which protein synthesis took place. These particles could be obtained from the cell using an ultracentrifuge. A clear looking but highly opalescent solution was obtained because it contained a lot of dissolved proteins. If this solution was taken and centrifuged at accelerations greater than 100,000g for an hour, a clear precipitate could be obtained at the bottom. This precipitate turned out

to be the particles that synthesise the protein. These particles are called ribosomes.
Although the opposite was assumed, it turned out that the composition of ribosomal RNA was different from the composition of cellular DNA, it also turned out to be almost identical in different species of organisms. That is, while DNA had specificity of nucleotide composition, ribosomal RNA did not.
Following Chargaff, because he had a small range of objects, but there were already indications that the DNA composition was species specific, several similar works were carried out, in particular by Davidson at the University of Chicago, and by Spirin and Belozersky at our department, who analysed the nucleotide composition of DNA and RNA in different organisms. Work with bacteria led to the fact that they found that in different species of bacteria the composition varies within very wide limits, i.e. there are bacteria in which 2/3 of the nitrogenous bases are guanine and cytosine and 1/3 adenine and vice versa, i.e. the variations are very wide, but the total RNA of the cell has no such variations, it is approximately the same. The point is that if we compare the composition of DNA and RNA of organisms, it turns out that, although all organisms have similar RNA composition, it is necessary to take into account the correlation between the composition of RNA and DNA. And if a distribution is made, it will appear that for a certain %GC of DNA, %GC of RNA will be in a different dependent value (Fig. 8.4). From this correlation, it was concluded that there are two fractions of RNA in the cell. The first one is approximately the same in all, i.e. fulfils a common function, and the second one carries information and correlates with DNA in its composition.

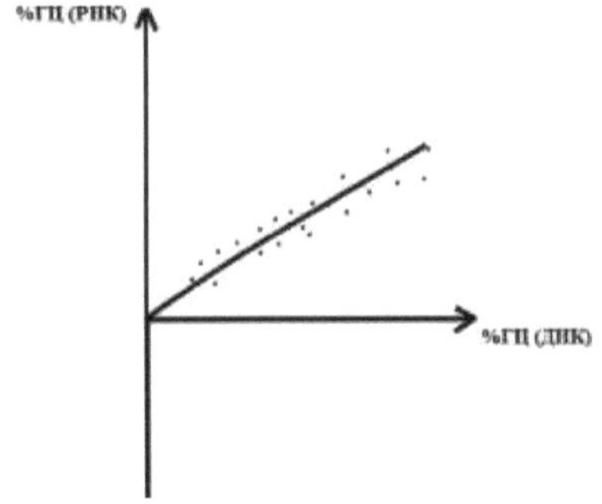

Figure 8.4 Correlation between guanine-cytosine percentage compositions of RNA and DNA

Following this, a group of American researchers carried out experimental work to produce RNA. They took E. coli cells and infected them with a certain virus. In a normal state, the cell synthesises different proteins and RNA, which should be similar to the DNA of the bacterium in composition and at the same time it should be bound to ribosomes. When a virus enters, it switches off the cellular systems and creates its own RNA and this RNA must go to the ribosomes and direct the synthesis of other proteins. The researchers gave radioactive phosphorus to the cell at the time of infection with the virus. The cells readily absorbed the radioactive phosphate and began incorporating it into nucleic acids and nucleotides. The tag was incorporated into ribosomes. Ribosomes were separated using sucrose gradient, i.e. sucrose concentration gradient was created in ultracentrifuge beaker, cell lysate was layered on top, then centrifuged and particles depending on their mass and size moved with different speeds, forming zones, then this gradient was divided into fractions and a certain ratio of components was measured in them, where UV absorbance was measured at 260 nanometres and it turned out that some amount of material was at the bottom, then a peak corresponding to ribosomes and a small amount of material on the meniscus was observed. At this point there appeared to be UV absorption outside of the infected cells. When particles of this peak were taken, it was found to be ribosomes.

In the course of research it turned out that there is a separate fraction of RNA that carries information from DNA to the site of protein synthesis, and ribosomes are the special structure for synthesis. Researchers have shown that there are two kinds of RNA, fundamentally different in their functions: one is ribosomal RNA, which forms its own particles in which synthesis takes place, the second kind is RNA that carries information (messenger RNA) - matrix RNA (mRNA).

The ribosome consists of two subunits, and each such subunit has one very long RNA at its base. The large subunit will have an RNA consisting of about 3500 to 4200 nucleotides. The small subunit has half as much RNA (somewhere between 1600 and 2000 nucleotides). There are two distinct groups: bacterial and prokaryotic ribosomes that differ in RNA size.

Ribosomal RNAs are not only similar in size but also very similar to each other in nucleotide sequence. This high conservativity of ribosomal RNAs is a standard in the field of systematics of living organisms. And somewhere since the 70s microbiologists have introduced the criterion of DNA nucleotide composition as a systematic feature, without which the description of a new microbe is not accepted.

The analysis revealed that substitutions in the small subunit occur in very small cases. And by the sequence of nucleotides it became clear that the whole kingdom of prokaryotes is divided into two distant branches.

Representatives of one branch live in rather extreme conditions and they survive in just such conditions where they adapted in ancient times. This group was named Archaebacteria, now they are simply called Archaea. It turned out that these organisms have a different structure of ribosomes, where there are additional areas that are also in eukaryotes, but not in bacteria. Some archaea do not have membrane phospholipids, their membranes are built of a different type of substance that allows them to live in areas where the temperature exceeds 200 degrees Celsius. This kind of thermophiles have another property - they are usually also oxidophiles. It turned out that according to the sequence of nucleotides in the small ribosomal RNA, they are very different from other bacteria, and very similar to eukaryotes.

Other types and functions of RNA

Another species that has been discovered in the study of protein synthesis are the so-called transport RNAs, which carry amino acids to the site of synthesis.

Thus, it should be noted that the main three RNA species have been considered, but there are many more. There are also RNA species that modify these RNA species at specific locations, that is, they determine the address and the mechanism of modification. There are more than a dozen classes of RNA in a eukaryotic cell and as a consequence they are all differently arranged and behave differently. But they are always single-stranded, and this is because each particular cellular RNA is a copy of a particular section of DNA, taken from only one strand. DNA diverges at a certain site, a complementary RNA is synthesised on one of the strands, and the second strand just waits for its turn to form a double-helix structure. Cells use this phenomenon to defend themselves against viral infections.

9. Transcription

Difference between RNA synthesis and DNA synthesis

RNA synthesis is called transcription. The fundamental difference is that in the case of DNA synthesis the entire molecule is copied, while in the case of RNA synthesis a certain section and only one strand is copied. The bond between the nucleoside triphosphate phosphate is carried out. The sugar carries a base that attaches to the matrix strand and fixation occurs at this site. If the pairing is correct, the phosphate goes to the third hydroxyl of the end of the newly synthesised filament. This all happens at the active centre, which has two firmly bound magnesium ions. These magnesium ions are coordinately and electrostatically bound to asparagic acid residues. Three of these residues hold the ions together. One is held quite firmly in complex with three residues, another is held with only two, and the third bonds to the pyrophosphate of the nucleotide. In this way there is charge compensation on the phosphate, which in turn attacks the oxygen at the third position of the ribose. Fixation by magnesium ions and sugar thus accelerates the course of the reaction. This is only possible if the nucleotide is complementary. The enzymes involved are called DNA-dependent RNA polymerases, and these enzymes were the first to be discovered in nucleic acid synthesis because RNA must be synthesised in large quantities at all times (Fig. 9.1).

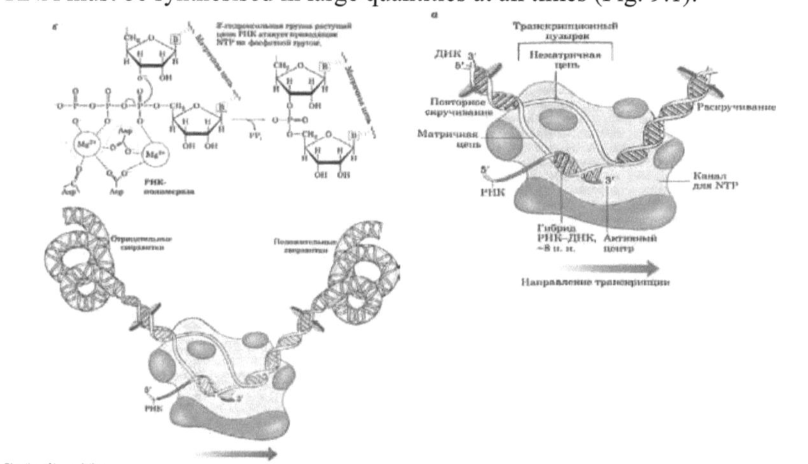

Figure 9.1. Mechanism of transcription

They differ from DNA polymerases in their properties. Unlike DNA-polymerases, DNA-dependent RNA polymerases do not need inoculum, this polymerase can independently form the first an inter-nucleotide bond. The second difference is that its errors are much more frequent (one per 10,000 cases on average). Also, the concept of processivity is not very applicable here, because the length of the resulting product depends not on the properties of the enzyme, but on the transcribed sequence. There are other sites present on the enzyme. Speaking of such enzymes, we should start with bacterial enzymes.

DNA-dependent PHK polymerase in bacteria

This enzyme, in addition to performing the polymerisation reaction, must be able to do a number of other things. Firstly, to make a molecule of a certain length: to start to start its synthesis at a certain in a certain place and end in a certain place. So it has to recognise those places. Firstly, it has an additional function of recognising the start of reading. Secondly, it must unravel DNA, because the matrix chain for transcription must be free from the complementary chain. Thirdly, it must return the DNA to its original structure after reading. In bacterial cells, a single enzyme does all this, and it has a complex structure. For starters, it is built from a large number of different subunits. It has a mass of approximately 350 kDa (Figure 9.2).

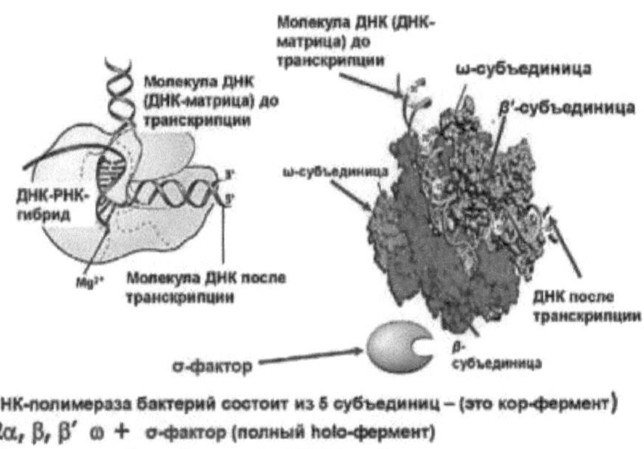

Figure 9.2. Structure of prokaryotic RNA polymerase

Enzyme subunits 2α, β, β', ω, σ (not always present). Often, the enzyme molecule is obtained without the last or last two subunits. If the enzyme is complete, the enzyme makes the correct RNAs on the matrices, and if the σ-subunits (38 kDa

= about 300 a/k residues) is not present, then it is less active, and also results in a set of fragments of random length. Thus, σ affects the accuracy of read start. That is, its function is to recognise the transcription start point. One of the groups of viruses on which studies were carried out was a group of T-even phages. These are rather large phages with a large (more than 50) number of genes. These genes are transcribed non-simultaneously. The RNA polymerase switches by reading each group (super early, early and late).

RNA polymerase was isolated from infected cells that had switched to synthesising on the late genes, and it turned out that when RNA was synthesised, only the late genes were synthesised. Same with the early genes. So it was as if RNA polymerase had two forms. But it turns out that viral RNA polymerase has a different subunit. That is, the virus changes the subunit in the cell, when it enters the cell in the early genes the bacterial (host) enzyme works, among the genes it transcribes there is a gene encoding its own σ-subunit. Once this subunit starts to form, late genes start to be read out. Among them is a protein gene that modifies and results in the degradation of the host subunit. Therefore, the early and host genes stop working and switch to providing the virus through its transcription on the gene. This mechanism takes place not only in viral infection, but also in the normal life activity of bacteria. It turned out that in addition to E. coli, there are several minor subunits in addition to the major subunit, which are responsible for specific groups of genes. One of them is associated with certain genes of nitrogen metabolism, others - with the synthesis of heat shock proteins (response to adverse influences), and these subunits recognise the beginning of different genes. What exactly is being recognised? Jacob and Manot studied mutations in E. coli for their ability to utilise lactose. It turned out that in front of the gene that encodes the lactose metabolising enzyme there is a non-coding region, but mutations in which cause the enzyme to stop being formed. They hypothesised that this region serves to attach an RNA polymerase there and named this region the promoter. The coding part usually starts at adenine, and in all the preceding regions there are two sequences that are somewhat similar to each other and sometimes completely overlapping. When these sequences were averaged, the sequence 6 nucleotides away from the beginning was TATAAT. The sequence averaged in this way is called the consensus sequence (Fig. 9.3).

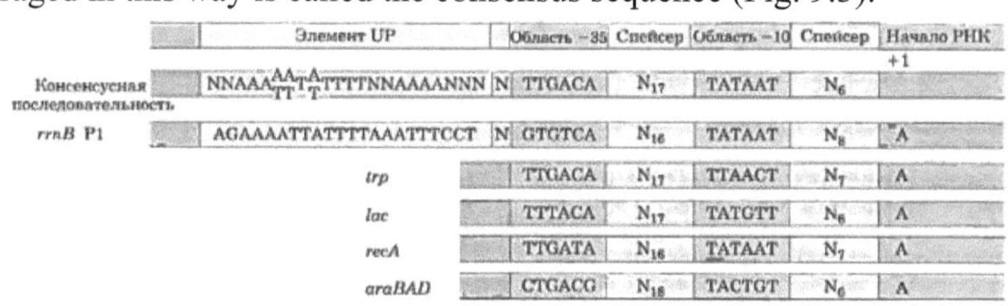

Figure 9.3. Typical E. coli promoters recognised by the RNA polymerase holoenzyme containing σ70. The consensus sequence for E. coli PROMOTERS recognised byis shown second from top. The intermediate sites (spacers) have a weakly divergent number of nucleotides (N). Only the first nucleotide of the sequence encoding the transcript (position +1) is shown.

If we take a sequence of 6 nucleotides and represent it by its geometric centre, it will be the 10th position from the beginning of synthesis, so this region was named "-10" region. Further on, after some distance (16-17 bp) there is another region, its centre falls on the position "-35", hence its name, and here also TTGACA is a consensus sequence. The closer the sequence is to the consensus sequence, the better it is recognised by RNA polymerase and the more often RNA polymerase starts synthesis on it. If the sequence is good (as in ribosomal genes), then a large number of products will be synthesised on it, if it is worse (as in the RecA protein), then a smaller number will be synthesised. Lac - a group of genes that encode lactose digestion, it will be even less due to the greater difference from the consensus sequences. araBAD - these are proteins of arabinose sugar digestion, and only three proteins are encoded in the region "-35", N between the regions is greater - this group of genes will work even worse.

In the same work, Jacob and Mano showed that transcription goes not of a single gene, but of a group of genes. That is, polycistronic matrices are formed, which are related to lactose metabolism, and the Lac operon contains three proteins, that is, it contains 3 proteins on which one mRNA is formed. It is worth noting that in the spacer between -10 and -35 sequences stands 16 or 17 nucleotides, however, when increasing to 18-19 nucleotides, the work stops, although this distance is small (only 7 angstroms). This is due to the fact that there is a helix shift! A, come to think of it, the β-subunit overlaps one side.

Since the consensus sequence is known, when constructing a gene or a group of genes for expression in 1 cells, one can simply put the most consensus sequence upstream. More often, such a fragment is synthesised and ligated to the corresponding DNA via restrictidase or ligase.

Mechanism of transcription

The enzyme, having bound the σ-subunit to this site, contacts not only the σ-subunit, the σ-subunit holds the enzyme by recognising the sequence. But the β and β'-subunits are arranged in such a way that the DNA is sort of sandwiched between them. That is, it goes into the gap between these subunits and is fixed there. Then there is a delay in synthesis, which is due to two processes. The first is that polymerase, having sat on DNA, sits on double-helical DNA and therefore cannot start synthesis because the helix is closed. And, although it can start synthesis on its own, it cannot unravel DNA on its own, so it sits and waits for fluctuations in the structure (it should be remembered that the interactions holding the double helix together are quite weak at each particular point). As soon as the strands have unravelled due to thermal motion, the polymerase fixes this section, forming a single-strand break inside (Fig. 9.4).

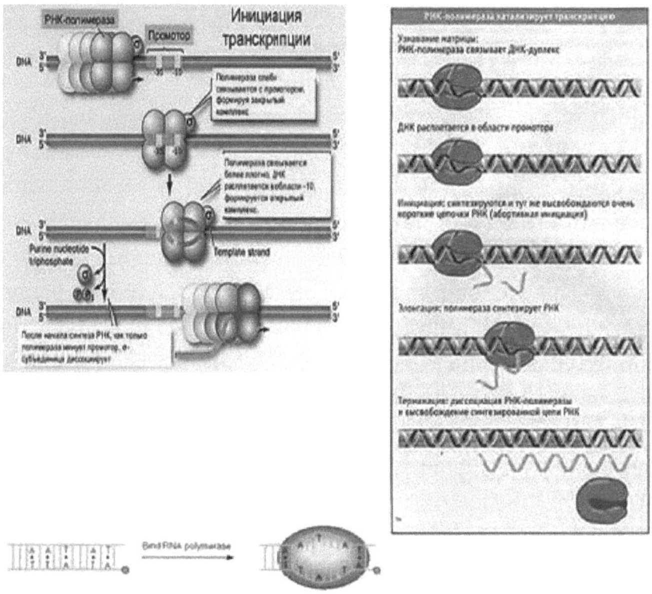

Figure 9.4. Transcription initiation (top), transcription mechanism (^paβa), and OPEN complex formation (BOTTOM)

This single-stranded site within the enzyme is small (6-8 stranded nucleotides). Once fixed, the enzyme begins synthesis at this site of RNA. RNA polymerase synthesises RNA strands and discards them if they are up to 10 nucleotides long (abortive initiation stage), after which it begins full synthesis of the long RNA strand. Presumably, this is because the RNA polymerase is large, and the RNA that starts to separate from the matrix during further synthesis has to get into a certain channel to get out. It doesn't always get in. If it gets stuck, it gets thrown out. Synthesis starts when the RNA gets into a certain channel, then the RNA starts to come out and the enzyme, moving along the matrix, continues synthesis by unravelling the RNA at the front end and braiding the DNA at the back end and separating the DNA from the RNA. The DNA-RNA helix is stronger than DNA-DNA, so energy must be expended to separate it. The energy source is the binding energy between the NTPs (nucleoside triphosphates) that enter the reaction and the phosphodiester bond that comes out of the RNA. The σ-subunit dissociates after synthesis has begun. It is worth noting that the polymerase does not unwind the DNA but superhelicalises it. Therefore, the presence of negative DNA superhelicalisation dramatically increases the rate of transcription initiation. The rate can be increased (for example, if the DNA is linear rather than helical) by adding topoisomerase II.

Termination

Polymerase has no preference for specific sequences, but there are areas where it slows down. These are called transcription pauses, and it's still not explained why this happens. It's not a stoppage, it's exactly a slowing down. When the enzyme passes this site, it moves until it meets a signal that signifies the end of this unit of transcription, called a terminator. Normally there is a special sequence at which synthesis stops, it has certain characteristics. This section contains an internal inverted repeat (palindrome), which causes the RNA to bend to form a hairpin (inside the polymerase) as it passes through this section. On the one hand the hairpin inhibits the polymerase, on the other hand it shortens the outer part, this reduces the speed of movement of free RNA and the polymerase itself. Secondly, after this hairpin there is usually an AT-rich DNA strand, and it is asymmetrical. That is, one strand - the matrix strand - has more adenines, so many AU pairs are formed. These pairs deform the helices, the bond between the matrix and synthesised RNA is weakened, which leads to the release of RNA from the complex. This is one variant of termination (Fig. 9.5).

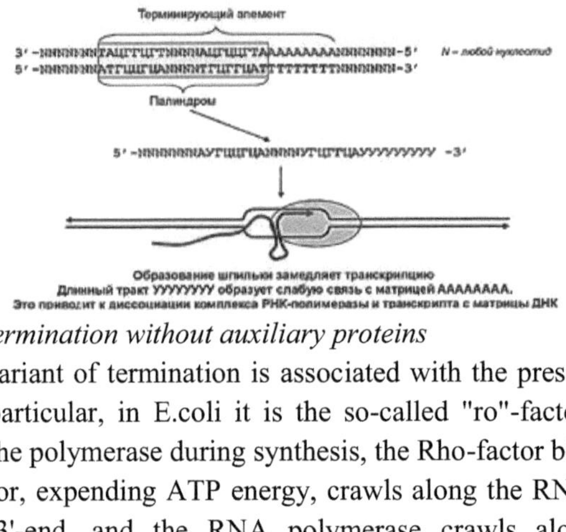

Figure 9.5. Termination without auxiliary proteins

The second variant of termination is associated with the presence of special proteins, in particular, in E.coli it is the so-called "ro"-factor. As soon as RNA leaves the polymerase during synthesis, the Rho-factor binds to it.

The Rho factor, expending ATP energy, crawls along the RNA from the 5'-end to the 3'-end, and the RNA polymerase crawls along the DNA, synthesising the RNA and lengthening the 3'-end. This continues until the polymerase gets stuck for some reason. Then the Rho factor starts pulling the RNA out of the polymerase and the complex breaks down. As a result, the RNA is released (Figure 9.6).

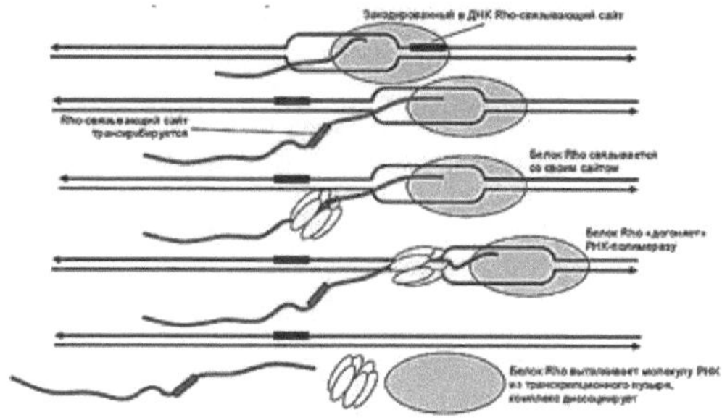

Figure 9.6. Termination involving auxiliary proteins

Regulation of transcription

After termination, everything goes back to the original state: the DNA is helical, the enzyme is released, and everything is ready to be repeated. But polymerase does not necessarily sit on the same promoter. Therefore, there are mechanisms of transcription regulation. The first is different promoters (some sites have more or less frequent initiation). The second one investigates the effect of substrate availability on the enzyme that cleaves it. Jacob and Manot investigated the digestion of lactose by intestinal lactose.
by E. coli. E. coli lives in the large intestine. It feeds on what the human body has not been able to digest. E. coli got into a situation where there was milk sugar as a food source. But lactose is not absorbed by cells, so it has to be broken down, or rather it is absorbed by cells poorly, and once inside it is still not metabolised unless it is hydrolysed. The enzyme that splits galactose from glucose is beta-galactosidase. But this enzyme is only needed when lactose is ingested with food. So not very often. Therefore, E. coli does not normally produce this enzyme. If you take a culture of E. coli, put it on a normal medium containing glucose and amino acids, it will grow well, but if you destroy the cells and measure the activity of beta-galactosidase, it turns out that the activity is zero. If you add lactose to this medium, then destroy the cells and measure the activity, it turns out that there is a lot of beta-galactosidase. Moreover, it begins to form quite quickly. Jacob and Mano received mutants that were unable to grow on lactose and mapped these mutations by genetic methods. It turned out that the mutations fall into three groups. The first is the occurrence of amino acid substitutions in the gene, that is, mutations that change the protein. So they mapped the gene itself. It turned out to be a medium-sized gene. In front of the beta-galactosidase gene sequence was a second group of mutations. Mutations in it would cause the enzyme to stop being formed altogether. They assumed that transcription starts at this site and called it a promoter.

In the same region, separate mutations appeared, which led to the fact that beta-galactosidase began to be formed independently of the presence of galactose. That is, the region turned out to be heterogeneous (constitutive synthesis). The third group of mutations turned out to be located rather far from the main gene, and mutations in this region led, as a rule, to the fact that the enzyme also began to be synthesised permanently, although sometimes the opposite was true: the enzyme was not synthesised even in the presence of lactose. It was assumed that this last region was some protein regulating the gene's work and was named genoregulator; the region in front of the beta-galactosidase gene, which gave constant synthesis, was named operator. It was further hypothesised that the genoregulator encoded a specific protein that recognised the sequence of the operator site. This protein was named repressor protein. By landing on the operator, the repressor protein blocks DNA at that site and prevents RNA polymerase from attaching to the promoter. Therefore, when cells grow in the absence of lactose, the promoter is blocked and RNA polymerase cannot sit on the promoter - information is not read out and no protein is formed. If lactose enters the cell, it binds to the repressor protein, and the repressor protein, as a consequence, cannot bind to the operator site. Binding is forbidden, but the promoter is opened. RNA polymerase sits on the promoter and starts transcription. Matrix RNA is formed and a protein, beta-galactosidase, is read from it. The protein cleaves galactose, and when all the galactose is cleaved, the repressor is released and sits on the gene again. That is, the enzyme is formed only when there is lactose in the medium (Fig. 9.7).

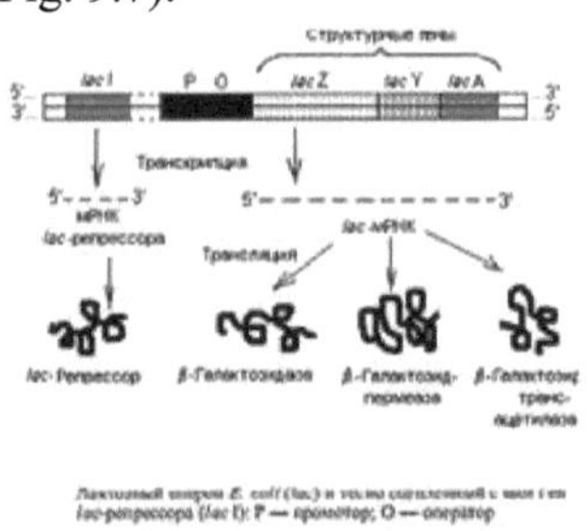

Figure 9.7. Regulation of transcription. Lac operon

There is a bottleneck in this scheme. In order for lactose to bind to the repressor, it must get inside the cell. It can't get inside the cell by itself. And it turned out that behind the beta-galactosidase gene there is a gene that encodes a protein that transports lactose across the membrane. It's called the lactose transporter. There is also a third gene that changes the type of bond in lactose, making it iso-lactose, which is a stronger ligand for the repressor protein. So when lactose gets into the cell, and there's quite little beta-galactosdase in the cell, this protein manages to isomerise some of the lactose, and it unblocks this operon. This was originally proposed as a scheme to explain genetic phenomena. Jacob and Manot also tested this on another E. coli gene, the alkaline phosphatase gene.

The regulation of alkaline phosphatase (ALP) synthesis is somewhat reversed. Alkaline phosphatase is needed to release phosphate from organics. Normally, this is a breakdown product in the environment of Escherichia coli. In order for the bound phosphate to become free and available for transport into the cell, it has to be hydrolysed, and alkaline phosphate does just that. It is located in the space between the cell wall and the membrane. Thus, only phosphate passes through the cell wall boundary and is immediately absorbed by the cell wall. The cell needs alkaline phosphate only when there is no free phosphate in the medium. When there is phosphate, cells do not form alkaline phosphate, but when there is no phosphate, they do. The task is reversed, and the mechanism is the same: there is a repressor protein, it binds to phosphate and binds to sequence operators in this form, i.e. the operator is recognised by the protein in complex with the ligand.

There may be cases when transcription is interrupted prematurely. This is the so-called attenuator regulation. In particular, it works in E. coli in the case of regulation of the tryptophan operon (a group of genes that are responsible for the synthesis of the amino acid tryptophan). Tryptophan requires a small amount of tryptophan. RNA synthesis begins on the DNA matrix, and following RNA polymerase, not a Rho factor but a ribosome sits on the RNA and begins to synthesise the protein. Moreover, it synthesises a site in front of the desired protein. This site has complementary sequences. If there is enough tryptophan in the cell, and tryptophan is incorporated into the ribosome in time, the ribosome will go some distance and block the formation of hairpins on the RNA. If there is not enough tryptophan, the ribosome will slow down at that point and the RNA polymerase will go further, leading to the formation of another hairpin, the terminator hairpin. That is, termination will occur immediately after initiation. As a result, synthesis will cease (Fig. 9.8).

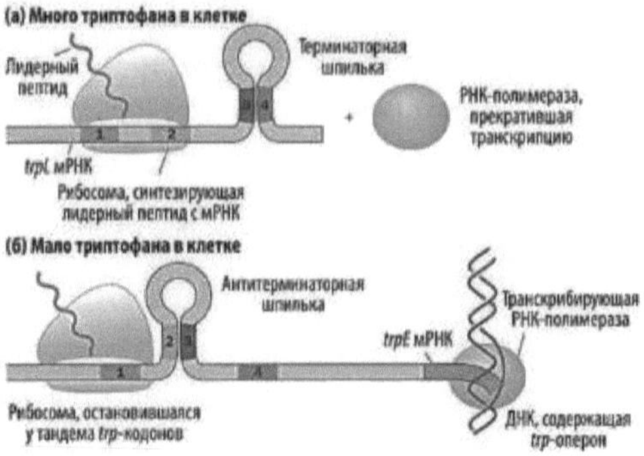

Figure 9.8. Attenuation of transcription

Transcription in bacteria

Synthesis of ribosomal RNAs in bacteria. Ribosomal, as well as matrix RNAs, are cistronic, i.e., a long RNA is read from a single promoter, which is further transformed into several different RNAs: three different RNAs that are part of the ribosome and one RNA that is a transport RNA (Fig. 9.9).

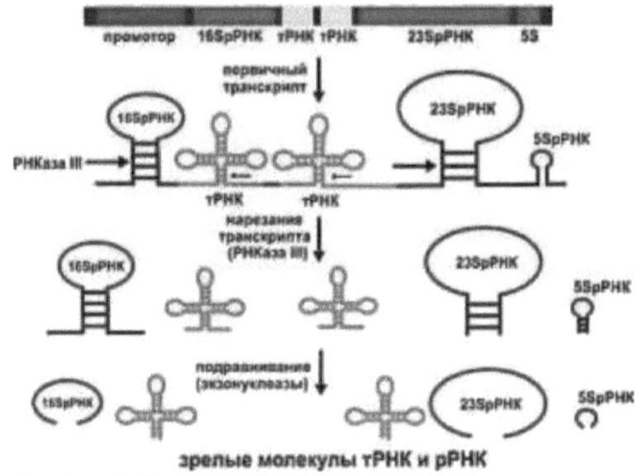

Figure 9.9. Synthesis of ribosomal RNA in bacteria

RNA from the position of individual genes is shown above. Certain regions of RNA are methylated (primary transcript), then endonucleases act to make an internal break, followed by 3'- and 5'-exonucleases, which detach nucleotides from the ends until the ends are complementary. In tRNA, the 3'-end is longer and contains 3 additional nucleotides (universal CCA-end).

10. Transcription and maturation of RNA in eukaryotes

Expanding on the theme of maturation of some RNA species, a figure (Fig. 10.1) is shown to illustrate the example of bacterial ribosomal RNAs, which form a single operon giving one initial RNA (at the top of the figure), which contains three precursors of ribosomal RNAs and one precursor of transport RNAs. The first step involves methylation using special enzymes to leave signals on the RNA that indicate its nature and boundaries. Next, cutting by specific nucleases takes place to form longer precursors. In the figure, the redundant sequences are shown in yellow. Further, all these RNAs have such a structure that their 5' end and 3' end are joined complementary to each other, a helical region is formed and the redundant regions are outside it.

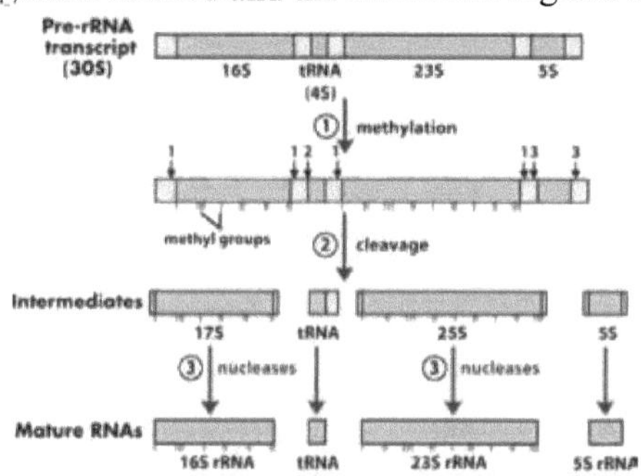

Puc.10.1 Biosynthesis of rRNAs and tRNAs

There are two special nucleases that remove these sites one nucleotide at a time. One of them acts from the 5' end to the 3' end and removes excess fragments (Figure 10.1 "MaShge RNAs"). These fragments are finished off by nucleases to the point where the helical secondary structure is formed. The remaining molecules (non-matrix RNAs) continue to function.

Because of the multiplicity of DNA polymerases in eukaryotes, the situation is more complicated. There are three completely different RNA polymerases in the nucleus of eukaryotic cells. They are labelled in the order in which they come off the chromatography column: first polymerase, second and third.

Principle of polymerase I

The first polymerase (Polymerase I) turned out to be an enzyme that synthesises only large ribosomal RNAs. In eukaryotes, these are usually named (based on size in animals) 18S and 28S RNAs, which forms the basis of the large and small subunit. Most eukaryotes have only these two genes in this operon, but there are exceptions. For example, in yeast, the operon also includes the gene for small ribosomal RNA (5S RNA). This is the only such case studied in eukaryotes, when the 5S gene is located in the same operon, but orientated in the opposite direction. Two genes are orientated in the same direction and read by one enzyme, and a third gene is in the same location but read in the opposite direction and by a different enzyme. The first polymerase recognises one kind of promoter.

Principle of polymerase II

The second RNA polymerase (Polymerase II) of eukaryotes synthesises mainly matrix RNAs, which carry information about the sequence of amino acids in proteins. This polymerase has to recognise a very large number of different promoters and work on them with different intensities to isolate the right genes to be active.

Principle of polymerase III

The third RNA polymerase (Polymerase III) differs in that it works on short genes whose products form many internal helices. From the point of view of the gene itself, this means that they contain many inverted repeats. It also mainly recognises its gene with the help of its auxiliary proteins, and these proteins bind, as a rule, on inverted repeats, i.e. there is no external promoter as such, but the promoter is the internal part of the gene, which encodes certain sections of the corresponding RNAs. First of all, polymerase III synthesises transport RNAs and 5S RNAs (small RNAs that are part of the large subunit of ribosomes).

Gene groups

Thus, the most complex variant is observed in the second polymerase. It turned out that there is another feature that complicates everything: the promoter region itself is fundamentally differently organised in some groups of genes. Speaking about genes encoding proteins and some other genes encoding RNA in eukaryotes in the case of complex multicellular organisms, it is necessary to pay attention to one more thing: in this case there are not two states of a gene (on and off, as in prokaryotes), but at least four.

The first group of genes was called housekeeping genes
(Housekeeping genes), and they are constantly active in all cells, i.e. their promoters are always in working condition, and they are responsible for basic metabolism. They are characterised by the fact that their promoter region (often the whole gene) does not fit into dense chromatin structures.

The second group is the exact opposite. In a given specific cell type, proteins are synthesised in some groups and not included in others. There are also intermediate variants, which consist in the fact that a given gene must produce a product in a given cell type in some specific amounts regulated by one or another factor. As a rule, such genes always show a small activity and in a small amount, they are read, their information is used for protein synthesis, as these proteins are needed all the time, only their amount is always changing.

The fourth group of genes is the most interesting, because these are genes that in most cases in this type of cells do not work, their products are not needed, but there may be special situations when this product is needed, and then this gene can potentially be switched on through some intermediate stages of activation. It goes from this fourth state to the third state by having specific proteins attached to it.

All these types are determined by the structure of chromatin in the promoter region, i.e. in those cases when these genes should not work. In this place the chromatin is densely packed, accordingly, the promoter region is closed, the protein factor RNA polymerase cannot join, and the gene is not switched on. Chromosome replication is arranged in such a way that when packing daughter strands, this structure is reproduced, i.e. those proteins sit there that lead to the fact that densely packed chromatin will appear in this place. When such genes form large enough areas, then what is visible by colouring in a colour microscope (so-called heterochromatin structures) is formed. If we talk about other groups of genes, then there the chromatin packing is not so dense, and the promoter region is not occupied by chromatin, but there are specific proteins sitting on it, which are ready for RNA polymerase to sit on them and start synthesis. In this case, the process will be limited by the rate of attachment of the RNA polymerase. In the case where a gene is working with a certain range of activity, then the proteins that are there will not be present in full, the rest of the

form a region of bound polymerase, but less efficient. In order for the gene to become more efficient, one regulatory protein must be added, in which case the polymerase will interact better and the activity will be greater. In cases where the gene can work but is in a dormant state, the promoter region contains certain transcription factors that prevent dense chromatin structures from forming at that location, but which are severely insufficient for polymerase attachment. The remaining factors that can join there are usually in response to some exogenous signals, and in this case genes that were not working are switched on precisely in response to certain signals.
The state of promoters in different cell types can vary. An initial attempt to identify promoters in eukaryotes was made in the same way as in prokaryotes in the work of Shain-Dalgarno, where they took the sequence of a small number of known genes and checked what they had upstream of the RNA.
Only one canonical sequence was found, similar to the -10 sequence in bacteria, only located at a distance of -30 nucleotides. That is, if the -10 sequence is a sequence one turn away from the start of synthesis, it is three turns away from the start of synthesis (Figure 10.2). The last two "A's" are optional, so this is usually called a TATA box (or pribnow box, pribnow box), which is a sequence that has something attached to it that triggers transcription.

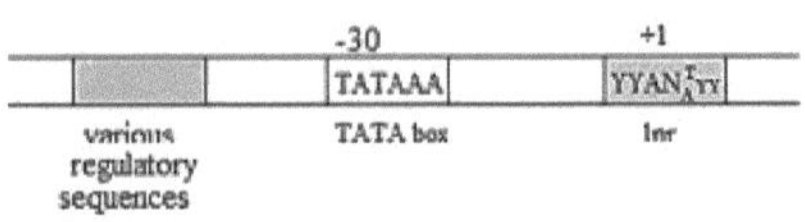

Figure 10.2 -30-sequence - TATA box

There is some consensus sequence at the site where synthesis starts (on the right, +1 is the point where synthesis starts). As you can see there are a lot of pyrimidines there, which are labelled with a "Y", and other than that there is nothing characteristic. At some distance after the -30 box, most genes have been found to have different sequences where regulatory proteins bind. Quite a few of their species are known and there may be several such sequences in front of one gene 112

dozens that may be involved in gene regulation.
In addition, another interesting structural element was found in eukaryotic genes, which was named enhancer. It is a sequence that is located absolutely anywhere (a distance of 1000 nucleotides before or after the gene, or somewhere in the middle), orientated in any direction, but enhances the work of a given gene. Unlike the sequences shown in Fig. 10.2, this one is not bound by physical contact with the gene, so we began to distinguish two groups of regulatory factors: the first is cis-acting factors, which have binding sites near the promoter, and the second is trans-acting factors, which are located far away and are not orientated in a certain way. It turned out that certain proteins bind to enhancers. Larger studies have found that there is a specific region where a given gene is located within which an enhancer can operate. In the same places where its action stops specific sequences have been identified. That is, there are specific sequences that restrict enhancers, which have been called insulators. These insulator sequences bind to the protein core of the chromosome, forming a loop in which a given gene, or group of genes, falls within the limit of the enhancer's action.
In addition to enhancers, there are sequences with the opposite effect, but with approximately the same properties. That is, sequences that are located within the same loop, regardless of orientation and position, will interfere with the work of the gene. They are called silencers (zyepseg). In some cells they work and the corresponding gene will be packed tightly and will not contain transcription factors on the promoter, while in other cells it will be switched off, because it will either get into a neighbouring loop during chromatin reformatting in specialised cells, or the necessary protein for its work will not be formed.
Later it turned out that there are cases that do not fit this scheme at all. It turned out that there are genes that do not have TATA boh, and the sequence in this region is not similar at all to that in all the others. Moreover, these turned out to be mostly housekeeping genes that work all the time. Despite its specificity, it still contains sites for binding certain regulatory factors. First of all, there are always GC islands there.
These sites turned out to be frequently methylated in DNA in these regions. This methylation had a certain effect on the work of genes and it turned out that it is just related to the fact that either methylated sequences bind the necessary factor, or vice versa - only unmethylated ones, and methylation performs switching. All these chromatin changes interact with RNA polymerase II.

Structure of polymerase II

RNA polymerase II is the largest RNA polymerase, it has two large subunits where one is almost always of a fairly defined size (~140,000 Dalton by mass) and has a characteristic binding structure that interacts irreversibly with the pale grebe toxin (alpha-mannitin). This is a probe for the second polymerase, as every molecule of this enzyme has this subunit and each such subunit binds alpha-mannitin so strongly that when the protein is denatured by detergents, this site remains unbroken.

Another large subunit is larger than this one (mass 220,000-240,000 Dalton). It was found to have features: it is present as several forms of different lengths. The ultimate variant of this difference was a sequence that had a mass of 180,000 Dalton at which activity could be measured. On further analysis, it turned out that this was not a natural state, and this piece is exposed and easily broken down, while being constructed from repeating units. It has an eight amino acid sequence that repeats several dozen times, and it contains many of these hydroxamic acids (like serine) and forms a hydrogen bonding system, coiling into a structure that protrudes from the overall protein.

In addition to these two large subunits, the polymerase includes more than half a dozen smaller subunits (with a mass of 60,000 Dalton or less). In general, the result is a very large molecule with a molecular mass somewhere between 500,000 and 800,000 Dalton. These complex molecules carry out the synthesis of informative RNA starting from the moment of binding to the promoter. It should be noted that the stable product of polymerase II has two other characteristic features at the ends, when matrix RNA is isolated from cells. It has a structure at the 5' end that is an inverted guanine nucleotide, which is called a cap (cap - hat). It is a guard at the 5' end, modified nucleotide and attached backwards by the 5' end to the 5' end. At the 3' end of the mRNA there has always been a long sequence made up of only adenine nucleotides. This sequence as well as the cap is attached after synthesis. That is, in addition to polymerase, a system that forms the cap and forms the poly(A) sequence - the tail (poly(A)-tail) - is involved in the formation of mRNA.

The tail is usually long (200 - 400 nucleotides) and it is still unknown what it is for. The poly(A)-tail is a way of identifying mRNA. An insoluble matrix is created and a sequence constructed only of thymine nucleotides is sewn to the cap, where a complementary site is formed, with poly(A) on one chain and poly(T) or poly(U) on the other.

Transcription of eukaryotes

RNA polymerase starts its work by searching for an attachment site. As mentioned earlier, there is a certain sequence (TATA box), which is present at position -30 in most genes, and certain proteins attach to it. It appears that the protein specifically binds to this sequence (thus the name TATA box-binding sequence) and by landing at this position, it creates a landing site for RNA polymerase. In Figure 10.3, it is labelled in the first position TWP (TATA binding protein) or TFIID (transcription factor 2 D). When it binds, other proteins additionally begin to attach to it, shown as separate circles in the figure. The RNA polymerase binds to a protein complex that helps the polymerase sit.

The result is a rather complex spatial configuration in which individual protein factors sit on their DNA sequences and at the same time other sites interact with each other, folding in a certain way and making it a convenient place for polymerase to land.

Next, just as in the case of prokaryotes the bound polymerase is over the double-helical DNA, after which the unwinding OCCURS, (which' in Figure 10.3 is a green loop), and the polymerase starts to move.

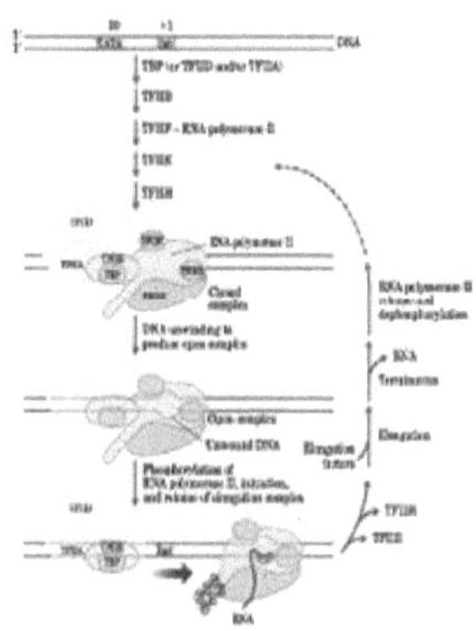

Figure 10.3 Transcription process in eukaryotes

Further polymerase, having sat on this place, as well as bacterial polymerase cannot immediately arrange the synthesised RNA in the right way, and therefore several short sequences are synthesised and discarded. In this state, the segment included in this initiator complex begins to phosphorylate the RNA polymerase. This protein kinase only phosphorylates serines in the repetitive sequences of the large subunit. Up to 30 phosphate residues are hinged on the enzyme. This is shown in the figure as a protruding part (the letter P in the brown circle) - a strongly negatively charged block. Additional components are attached to this block, which will further participate in the maturation process. The exit is performed in such a way that the end of the mRNA does not just fall into the medium somewhere, but falls on the phosphorylated end, to which a certain protein is bound, which immediately hangs a cap on the 5'end (Fig. 10.4).

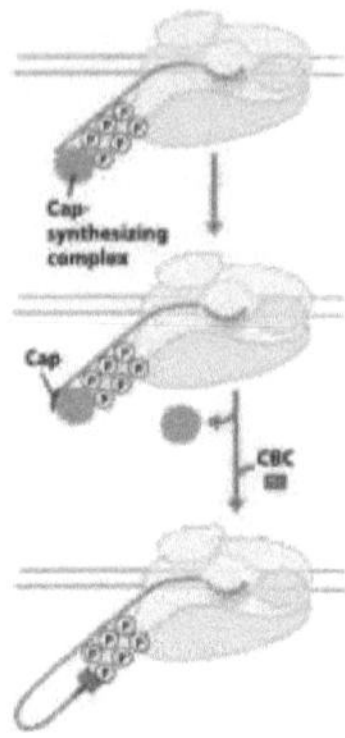

Figure 10.4 RNA caching

First of all, when the synthesis starts - the corresponding end is formed, there was a nucleotide containing triphosphate groups on it. So, the 5' end is a nucleotide with triphosphate groups. One of these phosphate groups is detached (diphosphate remains), then a GTP molecule comes in, and this molecule loses one phosphate, and with its second phosphate it attaches to the RNA phosphate, displacing the phosphate group, thus three phosphate groups are obtained to which ribose is attached on both sides - fifth positions. Then there is an additional modification, where guanine is methylated on the seventh atom, here a positive charge appears, and the first ribose in RNA is methylated on the second hydroxyl, i.e. the hydroxyl is blocked. This hydroxyl is important for certain groups of nucleases, that is actually this end is completely protected from normal nuclease degradation and has a characteristic grouping arrangement and is therefore recognised by certain proteins. This end is then fixed to the polymerase and the synthesis proceeds.
It turned out that in eukaryotes, matrix RNA does not contain the entire sequence of nucleotides present in the gene. That is, the DNA that encodes the given site is much longer than the RNA that is read from it. This reading process takes place over the entire molecule, but internal fragments are then cut out of it. If there is a DNA molecule, in the case of bacteria, the RNA is attached to it along its entire length. If we take eukaryotic RNA, it turns out that DNA is attached to it with some loops, which means that these sites are not in RNA, which was later confirmed genetically and these sites called

introns.

In some cases there are quite a lot of them, let us understand the chicken ovalbumin gene as an example (Fig. 10.5). Initially, the gene is given at the top and certain regions are shown in white and blue - regions that have different outcomes. RNA is synthesised with it, it has a cap at the 5' end and then certain sections of RNA are cut and removed and the rest are cross-linked together to form a shorter product from which the protein is synthesised. The final product will be just under 1900 nucleotides, meaning that a significant portion of the introns are removed. Eukaryotes have virtually no genes that do not contain introns. There can be up to several dozen of them, and their sizes are very different. It is not known exactly why they are needed, but there is a hypothesis that the removal of introns is a way of defence against the introduction of foreign DNA.

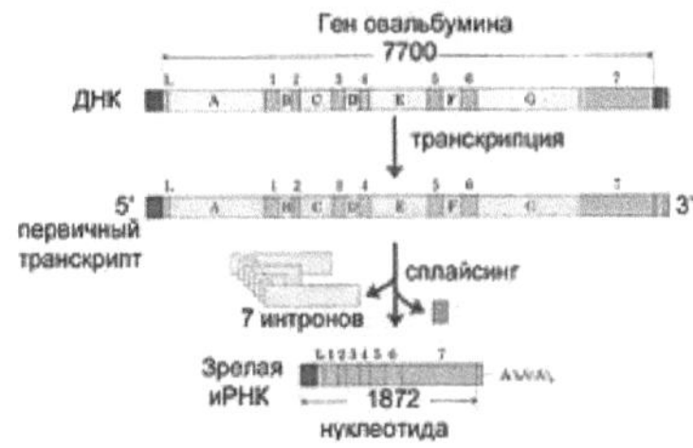

Figure 10.5 The chicken ovalbumin gene

The important thing is that mRNA loses internal sequences. The polymerase goes along the DNA matrix, synthesises RNA which has introns and exons and there is excision of introns and cross-linking of exon ends. Thus, a shorter product is made from a longer product. It turned out that in fact the movement of polymerase and the subsequent cutting of introns is also related to certain properties of polymerase.

Splicing

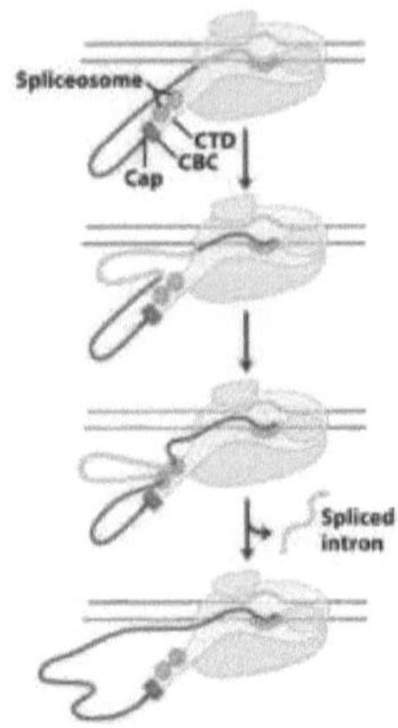

Figure 10.6 Splicing

In this case, a cap is fixed on the outgrowth, the rest of the RNA forms a loop and passes by another part of this protrusion, to which additional complexes are associated that carry out intron excision (Fig. 10.6). The excision of introns is called splicing. A complex system of structures is involved in this process: small nuclear RNAs (meRNAs, because they are rich in uracil) are involved, which recognise regions at the level of intron boundaries.

In the intron there is always a GU dinucleotide on the boundary on one side and an AG on the other, the important thing is that there are no universal sequences around them. Then the meRNAs come into play, where initially there is a complex of proteins with them, which binds ATP, in a certain way due to its energy is organised, and further proteins that are part of the complex holds dinucleotides, and the complex of two meRNAs, due to the fact that they have a lot of uracil nonspecifically forms a double-helix structure between adjacent sites. There is a guanine at this end, which is fixed so that it attacks one of the adenines in the middle of the intron (closer to the opposite end) and binds to it under the catalytic activities of this complex. This results in the formation of a lasso. At this point there is a change in the different steps of the different RNAs, and the loop is no longer held in the previous state and goes away, and the AG dinucleotide that ends the intron interacts with the OH-group, which interacts with the guanine bond with the next nucleotide and displaces the guanine, forming a continuous RNA from which the intron is excised, and the remaining lasso with adenine, which has 3 bonds with the AG at the end.

There is a phenomenon called alternative splicing, where not all introns are excised and not all exons are spliced. An example of this can be seen in Figure 10.7. There is an initial human hormone peptide gene, it has specific exons (shown in green) and introns (yellow) and different colour coded exons that denote the sequences encoding the corresponding peptide hormones. Two slicing steps are shown here: different splicing occurs in the two cell types (brain and thyroid).

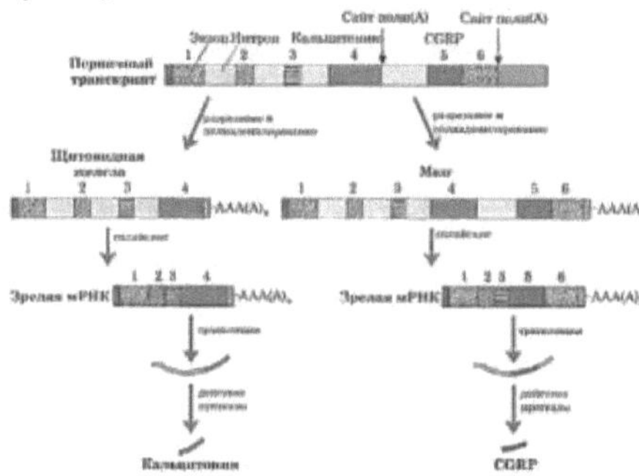

Figure 10.7 Alternative (splicing) processing of the calcitonin gene transcript

The poly(A)-tail is worth considering. It is associated with the process of RNA synthesis termination. In the second polymerase genes there is no well-defined termination. When reading from DNA, the resulting mRNA is capped, spliced, and then there is a certain sequence, which is the universal AAUAA base sequence in RNA. This sequence is recognised by a special protein complex that sits on the phosphorylated end, and this complex has two functional subunits, the first of them recognises this sequence and binds to it, and in a certain place after it makes a break, further the part falls off, and we see that the 3'OH-end falls into another functional part of this complex - the poly(A)-tail is built up. It turns out that before the poly(A)-tail there is always this universal AAUAA sequence. Building up of poly(A)-tail is a statistical process, and there is no clear boundary in it, after some time this complex falls apart, as these are not matrix polymerases with high processivity, and the corresponding poly(A) end is released, and further mRNA is ready to work. Only in this form of RNA - with a cap at one end and poly(A) at the other and introns cut out - can it leave the nucleus. For this purpose, there is a special complex in the nucleus - nuclear pores, through which large molecules pass, and mRNA is straightened, goes in the form of a stretched chain through the pore, and after it is folded again, although in some cases RNA goes in a complex with proteins.

11. Genetic code

Matching

Last time we considered the processes of RNA formation in eukaryotic cells. The final stage is the process of protein synthesis (translation). The difference between protein synthesis and other biopolymers is that proteins are irregular molecules with a strictly defined sequence of monomers. The difficulty in synthesis is to establish this sequence of monomers. In almost all cases of synthesis, the carboxyl group is activated. Activation can take place by the addition of a phosphate. As in the case of nucleic acids, synthesis requires information that is taken from some kind of matrix molecule. Such a matrix is mRNA. An mRNA has a sequence of four nitrogenous bases, while a protein has 20 amino acids. The problem was to establish the correspondence.

The question arises as to how many nucleic acid elements should encode one amino acid. If one nucleotide corresponds to one amino acid, then, as in the case of DNA and RNA, there are only 4 variants. If you take a combination of two nucleotides, you get 16 - still not enough. Of 3, it's 64 - way more than you need. Speaking of coding, Morse code comes to mind. Each extra sign, if it is repeated many times in the message, requires a certain time and cost, respectively it is profitable to compact it. In Morse code 3 signs encode a certain set of letters, there can be only 8 of them - the most frequent letters of the alphabet. If you take 4 each, you get 16 characters, whence 16+8=24 characters - enough for some alphabets. But usually not, so there are also 5-digit characters. The difference of Morse code is that there are boundaries (long pauses) between words, perceived by ear. In this case, however, the number of characters encoding an amino acid is the same for all amino acids so that the same blocks can be read. This can only be a triplet system (amino acids 20, combination of two nucleotides = 16, of three =64).

The next question is whether all combinations of nucleotides encode amino acids. Here there are variants: only 20 triplets encode 20 amino acids, and the rest are meaningless, the second - all amino acids are encoded by all triplets, then there will be more than one triplet for an amino acid. From the point of view of genetics, the first option was not suitable. The use of all triplets led to synthesis of 122

of a shortened protein that was rapidly degraded. Such mutations gave certain disorders precisely because these mutations resulted in the formation of triplets that did not code for amino acids. They were first called meaningless triplets, then it turned out that it is a triplet that means the end of the polypeptide chain. They were called stop codons. There are only three of them out of 64, and 61 encode amino acids.

The next question was how amino acids are defined in space. In the case of nucleic acids, it is hydrogen bonding and formation of pairs of the same size. If we compare amino acids with nucleotides, there is nothing like that here. There is a dimensional contradiction. If you look at the size of a nucleotide along the nucleic acid chain, it is almost three times larger than the amino acid fragment that is part of the polypeptide chain, that is, the nucleotide groupings are larger than the three alpha-carbon, nitrogen, and carbonyl atoms that make up the peptide chain. There are three nucleotides. This means that some amino acids will not overlap them because they are shorter in size. The assumption then arose that the triplets encoding the amino acids overlap (Fig. 11.1).

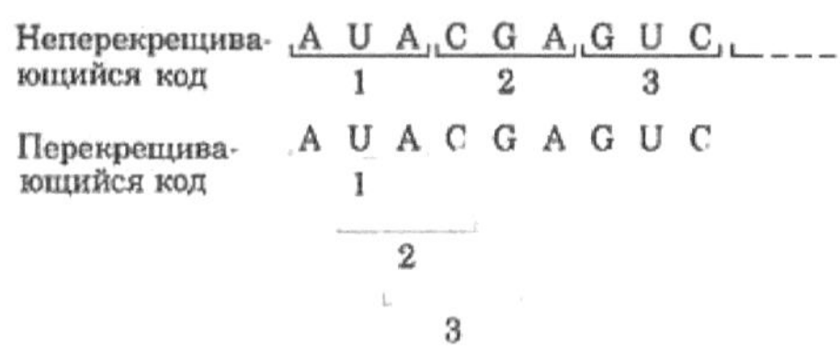

Non-crossing code
Crossed code

Fig. 11.1 Variants of triplets coding for amino acids

The objection is that in this way of coding the first amino acid strongly restricts the variants in the second position. That is, in such coding, this amino acid may be followed by some 4 amino acids. We did an experiment. They took yeast, poured acid and performed incomplete hydrolysis - the proteins broke down into different peptides containing different length sequences. We isolated a fraction consisting of two amino acids from there, isolated individual peptides from them and looked at their composition by complete hydrolysis and separation by chromatography. In this case, if one chemically modifies the amino group and, after hydrolysis, obtains one amino acid natural and the other modified, it is possible to determine 123 which one is on the N- end and which one is on the C- end. It turned out that all possible combinations occur. The code turned out to be non-overlapping.

How are triplets coding for different amino acids separated from each other? Are there any boundaries between them? From the chemical point of view, there are no such boundaries, because all nucleotides are organised according to the same principle. We can assume that one of the nucleotides is a boundary and the other three are coding. Then in the case of the triplet code there is no such a big redundancy (27 combinations for 20 amino acids), but on the other hand it led to the fact that in the sites coding for protein (at that time it was considered that all DNA codes for protein) in every fourth position must be the same nucleotide. That is, there is some periodicity of one nucleotide at a time. This has also been experimentally disproved.

Crick and Brenner's experiments

The clearest evidence of these properties of the genetic code was obtained in the experiments of Crick and Brenner. On the genetics of E. coli bacteriophages. E. coli has bacteriophages that destroy it. The viral particle attaches itself to the cell, injects its DNA into the cell, the DNA starts functioning there, viral proteins are synthesised, then the whole metabolism of the cell is switched to viral needs, and then a large number of copies of the viral genome are produced, and the cell is destroyed and the virus particles come out. This phage has an enzyme which, when the envelope is broken, dissolves it further, i.e. nothing remains of the bacterial cell wall, it breaks down into individual low molecular weight components. One particular gene is responsible for this. Crick and Brenner got mutations in this gene. What happened was the following: the virus developed normally in the bacterium, the virus killed the bacterium, the virus caused the destruction of its shell, but not dissolution. Therefore, where the virus was normal, there was nothing left of the bacterium - there was a transparent medium, where the virus was mutant - there was a turbid spot. Such viral plaques in case of mutation on this gene became turbid. Since the process is quite fast, it was possible to get a huge number of different crosses and see how inherited these or those features, even if they are extremely rare. Mutations were produced by the action of one class of compounds, acridine dyes. This is a structure of three conjugated heterocycles having two amino groups on sides (рис. 11.2). The amino groups create a positive charge that interacts well with phosphate, and these dyes bind well to DNA.

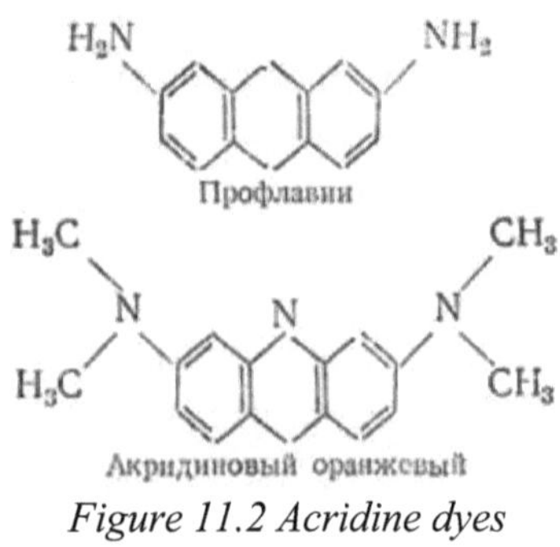

Figure 11.2 Acridine dyes

When cells are treated with this dye, they develop mutations that almost always result in the complete disappearance of the protein. Crick suggested that this dye is probably incorporated into the DNA molecule between the nitrogenous bases during replication (Fig. 11.3).

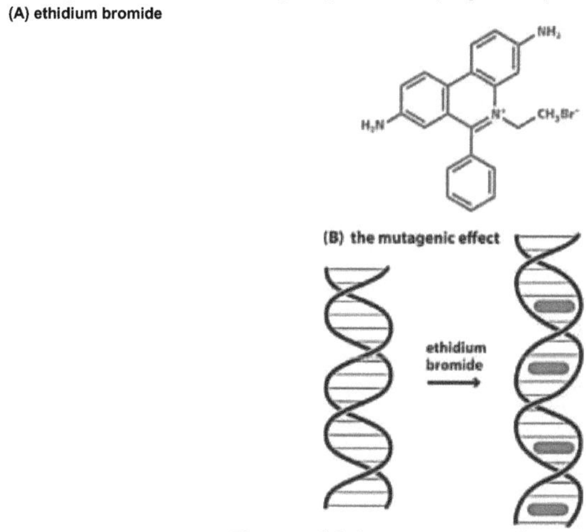

Figure 11.3: Dye intercalation in DNA

When replication begins, the insertion remains, and the polymerase makes an error at this point during synthesis. It was surmised that mutations cause either a single base to be dropped out or added. From here, a series of experiments began.

Phages were treated with dye and mutations giving turbid colonies were selected. Additionally, some colonies lacked the enzyme lysing cell walls. Next, bacterial cells were infected with a mixture of two such mutants. It turned out that in some combinations, infection with two viruses restored the ability to lysate cells. The assumption was made that, since in some cases there is a dropout and in others an insertion of a nucleotide, when crossing two viruses with two different mutations, the latter compensate for each other. Indeed, it turns out that all mutations can be divided into two groups. Within one group crossing does not give restoration of function, but when combining mutations from different groups with a fairly marked probability of such restoration occurred. This proved that there are no boundaries between codons.

It was later found that if the three mutations are not far from each other, recovery occurs quite often.

The experiments proved that the code was triplet ("the code is a multiple of three"), the code does not contain commas and does not overlap. The consequence was that most triplets encode amino acids (Fig. 11.4).

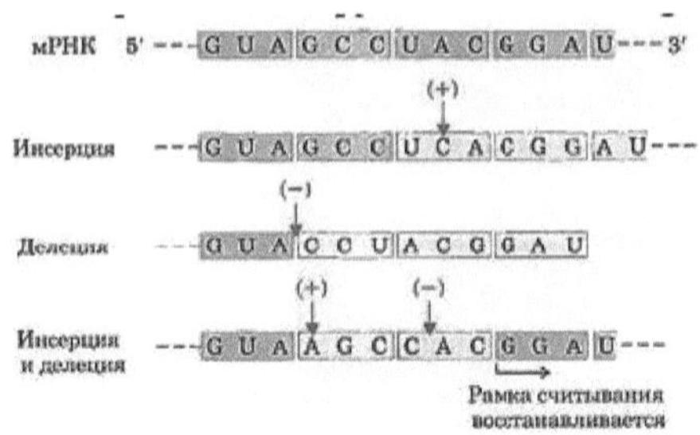

Inssrcia

Insertion and deletion
Reading frame
recover

Figure 11.4: Non-overlapping triplet code. The combination of insertion and deletion alters individual amino acids, but the result can restore the original amino acid sequence. The grey codons transcribed from the original gene are marked in grey; The blue colour shows new codons resulting from an insertion or deletion

The history of deciphering the genetic code

Practical interest continued to be aroused by the question: which acid is encoded by which nucleotide. While at that time it was already possible to decipher the sequence of amino acids, it was still unclear how to decipher the sequence of nucleotides. In the case of DNA, these are huge molecules where it is not clear where to start, in the case of RNA there was a problem of stability and heterogeneity.

Since the structure of mRNA was already known at that time and that protein synthesis takes place on ribosomes, some researchers created protein synthesis systems of a certain kind to investigate the mechanism in detail. Synthesis of a particular protein requires a particular mRNA, but it is almost impossible to obtain individual mRNA from cells. Then they took viruses, whose viral particles are RNA genomes, single-stranded RNA, from which protein is directly synthesised. And since there are viruses in which this RNA is quite small (for example, the tobacco mosaic virus has only three genes). The question arose as to how to separate the newly synthesised protein from those involved in the synthesis (auxiliary factors that are part of ribosomes). They decided to use labelled atoms (^{14}C or ^{3}H). But the added amino acids, besides being incorporated into polymers, can sorb on already existing proteins and nucleic acids. Control of the sorption background was needed. But no component could be removed. So, it is necessary to create conditions where sorption takes place, but synthesis does not. One of the American researchers Nirenberg as a control came up with a molecule - a polymer consisting of one uracil (poly-U). He introduced viral DNA into one sample, poly-U as an RNA mimic, added radioactive amino acids, the right components from the extract, incubated and measured the radioactivity. It turned out that the poly-U extract included radioactive amino acids. He surmised that the amino acid being included was the one encoded by the triplet containing three uracils. He then proceeded to take one radioactive acid at a time instead of a mixture and found that only phenylalanine was well incorporated, i.e., only one amino acid as expected. On the example of an artificial matrix, the first codon of the genetic code was deciphered.

There were also researchers who spliced nucleotides. These are Ochoa and Grunberg-Manago (she was his PhD student). She made a discovery that was interpreted as "discovering a way to synthesise RNA". That 127
so she got an enzyme that made RNA from nucleotides. From nucleotides containing two phosphate groups, i.e. ADP, UDP, CDP and GDF, from a mixture of which a polymer was made. The enzyme was isolated and characterised, but it turned out that this enzyme did not work on the matrix. It

also turned out that the polymer turns out to be statistical, that is, it randomly joins nucleotides in some sequence, and the frequency of inclusion of nucleotides depends on their concentration. As it turned out, it is an enzyme of RNA decay and has nothing to do with synthesis, but performs phosphorolysis of RNA (breaks the bond in RNA between phosphate and the third hydroxyl of ribose of the neighbouring nucleotide and attaches another phosphate group to the phosphate group at the place of the break). Polynucleotide phosphorylase proved to be a handy tool, but since Ochoa had this enzyme, he began to synthesise artificial matrices with different nucleotide ratios. First, he made Homopolymers. While poly-U gives phenylalanine, poly-A gives polylysine, poly-C gives polyproline, and poly-G gives nothing. Later it turned out that poly-G forms a multi-chain helix which the ribosome is unable to break down. But if a certain amount of adenine (80% G/ 20%A) is added to guanine, then no spirals are formed and it turned out to be glycine. By such variations it was possible to determine the composition (not the sequence) of triplets for each amino acid.

Nirenberg then began to use a binding rather than synthesis system, obtaining the binding of an amino acid using RNA on the ribosome with a triplet.

Koran was able to add synthetic results. He developed a method of cross-linking short sequences into long ones, for example, he cross-linked the dinucleotides AUAUAUAUAUAU and obtained triplet alternation. And he also determined which amino acid was encoded by which nucleotides using a radioactive tag. Thus, by 1966, the genetic year had been decoded.

Universal Scream Coding Dictionary

Crick identified a pattern: if an amino acid is encoded by two triplets, the first two nucleotides in these triplets are the same, and the third nucleotide is either one of two purines or two pyrimidines. Suppose for phenylalanine the codon is UUU and UUC, lysine: AAA and AAG. For an acid encoded by the three codons of the first two letters are the same, and 128 the third is any, except guanine. Many are encoded by four codons, with the first two letters being the same and the fourth letter being any. If six, then two according to the principle of four and two according to the principle of two (Fig. 11.5).

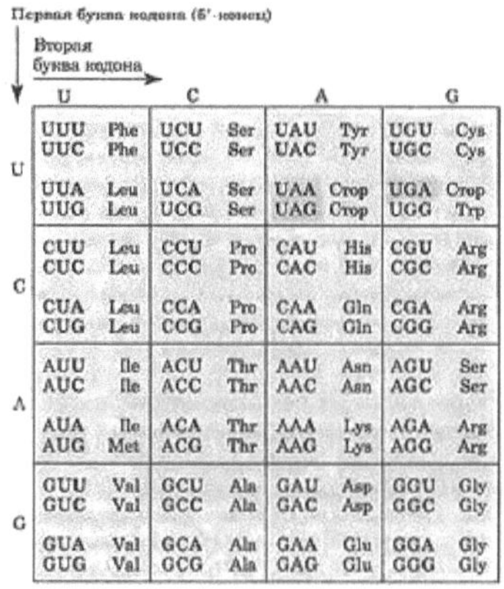

Первая буква кодона (5'-конец)

Вторая буква кодона

	U	C	A	G
U	UUU Phe UUC Phe UUA Leu UUG Leu	UCU Ser UCC Ser UCA Ser UCG Ser	UAU Tyr UAC Tyr UAA Стоп UAG Стоп	UGU Cys UGC Cys UGA Стоп UGG Trp
C	CUU Leu CUC Leu CUA Leu CUG Leu	CCU Pro CCC Pro CCA Pro CCG Pro	CAU His CAC His CAA Gln CAG Gln	CGU Arg CGC Arg CGA Arg CGG Arg
A	AUU Ile AUC Ile AUA Ile AUG Met	ACU Thr ACC Thr ACA Thr ACG Thr	AAU Asn AAC Asn AAA Lys AAG Lys	AGU Ser AGC Ser AGA Arg AGG Arg
G	GUU Val GUC Val GUA Val GUG Val	GCU Ala GCC Ala GCA Ala GCG Ala	GAU Asp GAC Asp GAA Glu GAG Glu	GGU Gly GGC Gly GGA Gly GGG Gly

Figure 11.5: "Dictionary" of the amino acid code in mRNA molecules. The codons are written in the 5' -> 3' direction. The third base of each codon (in bold) plays a less important role in the amino acid specification than the first two. The three stop codons are shown on a pink background, the initiator codon AUG on a green background.

Thanks to the universality of the genetic code, proteins encoded in the genome of one organism can be synthesised in the genome of another organism. This is how, for example, about 80% of all vaccines in the world are produced.

12. Transport RNAs

Today we will look at the processes that lead to the formation of the peptide bond in proteins. There is an RNA matrix that contains sequence information, a genetic code, which can be used to translate this sequence of nucleotides into a sequence of amino acids. For this process to proceed, each amino acid must be correlated with its codon. Here, spatial problems arise: the amino acid is large, but the nucleotides are three, and they are large. Watson and Crick proposed a solution to this problem. They suggested that there are adaptor RNA molecules that are complementary to the codon for a given amino acid (they have a site that can form a double-helix structure with the codon). On the other hand, they have a sequence to which a particular amino acid is specifically attached. Thus, the amino acid is connected to a sequence that can recognise the codon and bind to it.

It turned out that an amino acid, before being incorporated into a protein, is incorporated into RNA, with the incorporation of an individual amino acid that is not linked by peptide bonds to other amino acids. These RNAs were called transport RNAs or tRNAs.

Structure of tRNA

tRNA, although having different nucleotide sequences, has a common plan of structure: the 5' and 3' end sites are mutually complementary tRNA is shaped like a cloverleaf and consists of 4 short double-helical sections. These helical sites are connected by loops, which contain a lot of modified bases: dihydrouridine, thymine and pseudo-uridine, methylcytosine, dimethylguanine. Hence the name of the loops: dihydrouridyl (D-loop), thymidine-pseudouridyl (T- or Tψ-("T-psi")loop). The middle loop has three nucleotides complementary to the codon of a given amino acid (anticodon loop). At the 3'-end of the loop is a universal for all tRNAs

- CCA-sequences. They are attached to the prepared tRNA by an enzyme, not encoded (Fig. 12.1).

The spatial structure of tRNA was of interest. Despite being compact and structured by hydrodynamic parameters, RNA does not fit well into crystals. Then they started to use auxiliary compounds that bind to 130 phosphates and increase the affinity of the molecules for each other (detergents), but the crystals were unsuitable for PCA. The molecules were arranged regularly but were mobile. Then they took a reagent and modified nucleotides not involved in the formation of double helices. It turned out that loops are not always open to interaction, poorly amenable to modification. The reagents were cross-linked into a dimer with a methyl group in between. The reagent at each end looked for a different nucleotide, which allowed us to understand the environment of the groups. It appeared that the structure was additionally folded (Figure 12.1).

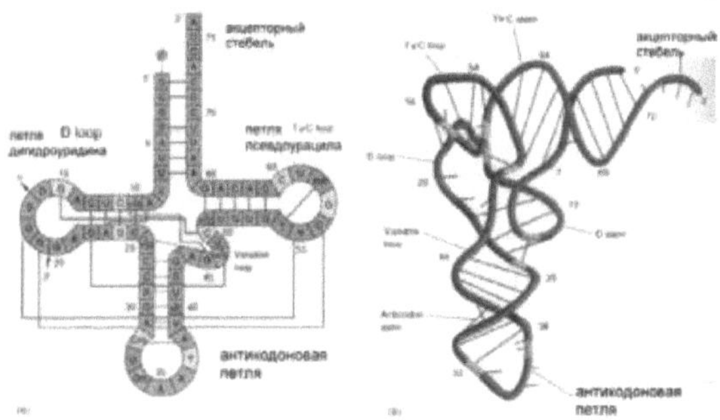

Fig. 12.1. tRNA structure

This produces not 4 helical sites but two, constructed each from two halves at right angles to each other, with an amino acid at one end and an anticodon at the other. The other two loops overlap, resulting in unusual interactions (Fig. 12.2).

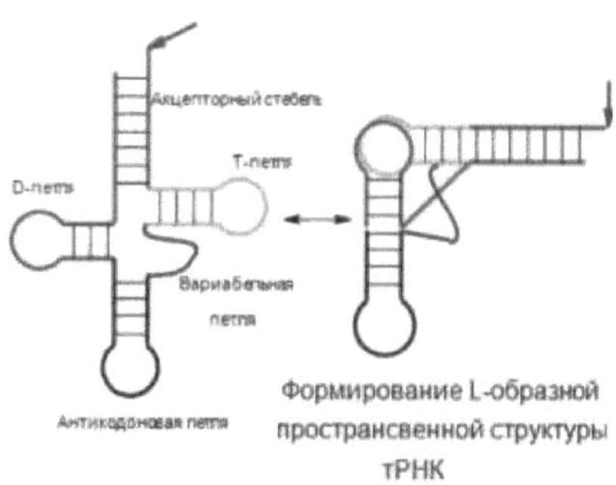

Fig. 12.2: Structure of tRNA spatial structure

Guanine triplets are formed (guanine from the opposite strand and a neighbouring one in the same strand), between which stacking occurs due to hydrophobic interactions, due to which the centre is well structured. Thus, there is an RNA globule from which two helix-strands emerge. The amino acid at one end is attached to the 3'-end to the ribose, and a chain break occurs at the anticodon.

Amino acid attachment to tRNA

The attachment of an amino acid to tRNA requires energy. The source of energy is ATP, and the enzymes that carry out the process are called aminoacyl-tRNA synthetases. The process is multistep. The first step is activation of the amino acid through the interaction of the amino acid with phosphate with the detachment of pyrophosphate. This bond is similar in nature to the bond in ATP, but it is not phosphoric anhydride, but mixed, is a greater energy source than ATP. The reaction is completely shifted to the product side due to hydrolysis of the pyrophosphate by the enzyme. The products are aminoacyladenylate. The tRNA binds to this enzyme, and the amino acid is transferred from aminoacyladenylate to tRNA when it approaches the 3'-end (Fig. 12.3).

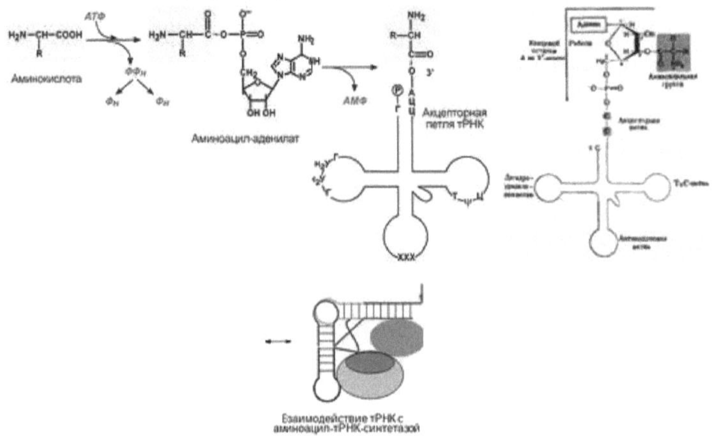

Figure 12.3: Amino acid attachment to tRNA

The reaction proceeds when there is a negative free energy shift. Since the anhydride bond is replaced by a ester bond, there is such a shift. Although the bond is ester bond, but another ribose hydroxyl is present nearby, so the energy is greater than that of the phosphodiester bond. Eventually the energy of ATP ends up in this bond and serves further to form the peptide bond. Attachments occur not only by 2- but also by 3-OH, but it is possible for an amino acid to move from one hydroxyl to another due to small distances and electrostatic interactions. That is, a constant equilibrium. The tRNA carrying the amino acid in the third position enters into further synthesis. Therefore, a protein comes into play that binds the aminoacyl-tRNAs after they separate from the enzyme that attaches the amino acids. This protein delivers the resulting complex to the ribosome and the 3'-attached amino acid gets there. There are two types of enzymes that attach amino acids to tRNA, which differ in a number of properties. The first property is the site of attachment: some attach to the 3'-end, others to the 2' end. From a spatial point of view, these enzymes attach amino acids from two different sides.

Second, the two types have different spatial stacking of polypeptide chains. In one group for adenyl nucleotide binding and amino acid attachment occurs according to the Rosmann stacking. Then the interaction between the radicals of the structures begins. Rosmann stacking is a standard variant of alpha helix stacking on beta layers, which leads to attachment at the 3'-position. Another variant is that ATP binding is formed by a flat beta structure with groups interacting with the adenyl ring, which leads to attachment at the 2'-position.

Thirdly, these enzymes are dimeric. Moreover, the dimeric nature is different - dimer and tetramer. The enzymes have different origins.

Attachment of tRNA to aminoacyl-tRNA synthetase

Binding takes place at many positions: interaction with the anticodon region as a rule, with the central part (where two loops overlap) and with the 3'-end (to participate in the reaction). In the end, there is a structure: an enzyme encompassed by a tRNA cap at both ends. Removal of the tRNA is poor, which is why the enzymes are dimers, as they work one at a time. The first binds ATP, attaches an amino acid to ATP, and transfers it to RNA, while the second binds ATP and attaches an amino acid and binds tRNA in the meantime. Since the tRNA is large, it collides with the first subunit and the tRNA on it, helping it get off the subunit (a type of ping pong), speeding up synthesis.

One would expect that if there is a different synthetase for each tRNA, there are as many synthetases as there are tRNAs, but this turned out not to be the case. The phenomenon of fewer tRNAs relative to codons was a problem. For each amino acid, the codons that are known for it were listed. Crick noticed that codons are similar, and if there are differences between codons for one amino acid, they are always in the third position. If there is one codon, the last codon in it will be guanine The exception was lysine - has three codons.

The hypothesis of ambiguous correspondence

Crick suggested that the last nucleotide of a codon and the first nucleotide of an anticodon interacting with it may in a non-standard way (Ambiguous Matching Hypothesis). He suggested that interaction with the formation of not strictly canonical but also similar pairs is possible (Fig. 12.4-12.5).

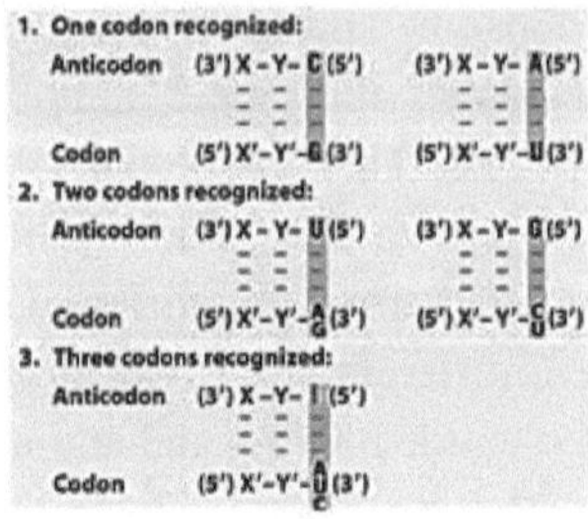

Figure 12.4: How unstable anticodon base pairs determine the number of recognisable Trnk codons

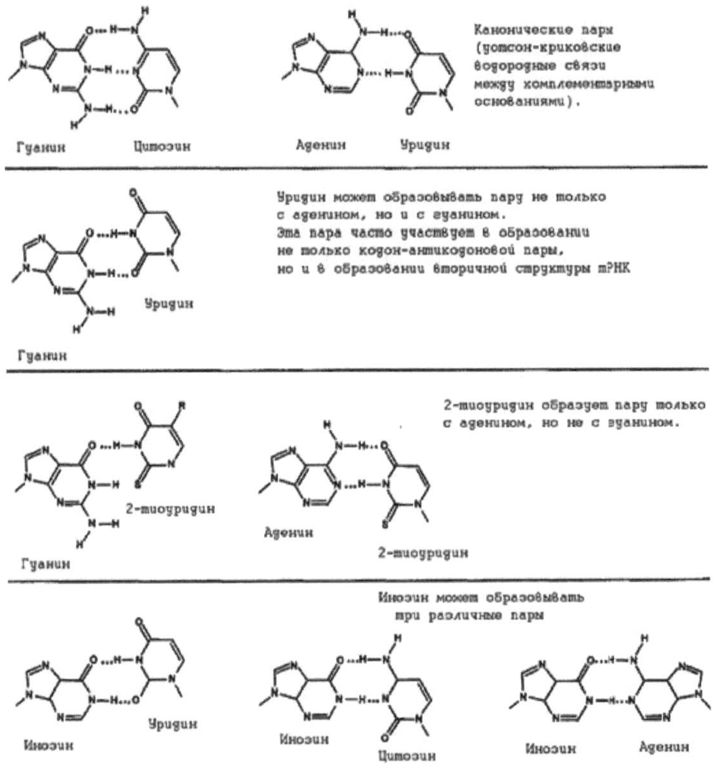

Figure 12.5: Ambiguous correspondence in codon-anticodon pair formation allows some bases in the first anticodon position to recognise more than one base in the third codon position

Quite a FEW ACIDS have 4 codons. How is this achieved? While a single tRNA can recognise 1, 2 or 3, 4 can be assembled by joining two twos, or a triplet and a unit. In most cases it is two twos. At the first position of the anticodon U, which recognises either A or G; or at the first position of G,

which recognises both U and CD. That is, in the case of proline, there may be a variant where there is a tRNA with the anticodon GGG, or c UGG (codon and anticodon are antiparallel, i.e. read in different directions).
There are some amino acids that have 6 codons, such as arginine. In this case there will be 3 tRNAs: one will recognise the two and the other two will recognise the four. Different codons recognised by different tRNAs are generally found with different frequencies. Accordingly, for those codons that occur frequently, the tRNAs that recognise them are also represented in large numbers. Those that are rarely represented, for them the corresponding tRNAs are found in small quantities. In the process of protein synthesis, if such a rare codon appears, there is not enough tRNA for it, and it takes a long time to find it; if there are several of them in a row, it is a noticeable pause in protein synthesis. It is possible that because of this, the protein starts to fold not quite correctly.
So, tRNA is ready to work: an amino acid is attached to it, energy is stored in this bond, it has its own anticodon, which can recognise the codon, it remains to bind it to mRNA so that the amino acids of these RNAs are brought close and joined together. The ribosome exists for this purpose. These are universal particles. According to their external parameters, they are divided into two groups: bacterial and eukaryotic. The sizes of ribosomes and their complexes are measured in Svedberg units (a coefficient describing the rate of deposition of a particle in some force field: gravitational or centrifugal), because ribosomes are easily separated by sedimentation.

CHAPTER 13

13. Ribosomes and systematics. Initiation of translation History of discovery of ribosomes. Ultracentrifugation

Within the initiation process, attention must be paid to the tRNA molecules carrying amino acids and the matrix RNAs with which they must interact. The name "ribosomes" is given to these particles because they are mainly composed of RNA. Ribosomes exist to organise a process in space so that all the components are in their proper place. Ribosomes were discovered long before scientists figured out their purpose, this discovery is due to the fact that ribosomes are clearly visible in an electron microscope. Ribosomes are usually identified by "staining" a cell slice with heavy metal cations and looking at them in an optical instrument.

These particles were found to be about the same size within the resolution of the microscope in all eukaryotes and all prokaryotes. In general, these particles have a unique size: they are much smaller than all cellular structures such as mitochondria, lysosomes, and chromatin. It is also possible to see ribosomes by centrifugation - they precipitate in a clear solution. Further studies were out by ultracentrifugation, thus obtaining certain values for the sedimentation rate. This method does not provide precise data, but helps to give a general characteristic of the behaviour of particles in viscous and aqueous media. As a conventional unit for carrying out the necessary calculations, scientists have identified the so-called "Svedberg unit" and it is denoted by the symbol "S". Thus, depending on the settling velocity of the particle in comparison with a certain standard, the relative values were deduced

valuesrate of sedimentation. These are the characteristics

behaviour of the particle in solution, which are not additive.

It turned out that measuring sedimentation rates for whole ribosomes, gives the following values: 80S in eukaryotes and 70S in prokaryotes.

Effect of Mg+. NH+, K+ and polyamines on ribosome stability

Further it was found that in phosphate buffers these particles begin to disintegrate. Subsequently, buffers were replaced by other solutions that stabilised the particles, but the stability was still low until magnesium ions were introduced into the medium. Magnesium ions were introduced a lot - magnesium accounts for 1-2% of the total mass of the ribosome. This means that almost all negative phosphate charges in RNA ribosomes are compensated by positive magnesium charges. Magnesium actually forms inter-strand cross-links because it is divalent and binds to two phosphates, and if it is taken away, the ribosomes dissociate. If you reduce the concentration of Mg to 1 mMol for prokaryotes, and slightly more than 0 for eukaryotes, the ribosome does not completely dissociate, but splits into two subunits: (eukaryote 80S^ 60S+40S; prokaryote 70S^ 50S+30S). The subunits can be obtained in pure form by centrifugation, and if the resulting individual subunits are mixed together and Mg is added, they are converted back into a whole ribosome, i.e. this process is completely reversible. Schematically, this can be depicted as follows:

$$\textit{Eukaryote}\ 80S \leftrightharpoons 60S+40S$$

$$\textit{Prokaryote}\ 70S \leftrightharpoons 50S+30S$$

Isolated ribosomes, upon addition of certain components or ribosome-free cell lysate of the amino acid ATP, are capable of synthesising peptide bonds. Potassium ions (K^+) serve as a stabiliser for ribosomes, while sodium ions (Na^+) are the opposite. This explained the fact that potassium ions are abundant and sodium ions are scarce in cells. The average concentration in an animal or bacterial (e.g., E. coli) cell of K^+ is 150-200 mMol and No' is 10-15 mMol.

In addition to Mg, ribosomes include polyamines such as spermine, spermidine - these are substances synthesised from amino acids by removing carboxyl groups and joining side radicals carrying amino groups such as carnitine with residues of other amino acids. This produces a molecule containing 4 to 10 carbon atoms (C) to which 3 or 4 amino groups are bonded - the result is a positively charged organic cation that is incorporated as counterions into ribosomes.

Analysis of rRNA sequences

If the ribosome is treated with a reagent that removes the protein, virtually pure RNA remains because of the fragility of the bonds of the low molecular weight components. These RNAs have been shown to be the major RNAs observed in the cell. Ribosomal RNAs account for about 80% of all cellular RNA. After the emergence of methods allowing the determination of primary RNA structures, following transport RNAs, attempts to determine the nucleotide sequence in ribosomal RNAs began.

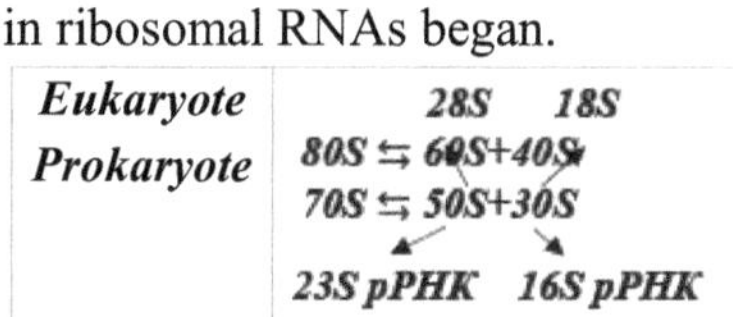

At this time, researchers developed methods for rapid DNA sequencing. The discovery was made that it was possible not to sequence RNA, but to isolate tRNA genes from a particular type of organism's cells and read it. As this method proved to be more cost-effective and faster, it began to be used on cells not only of

Escherichia coli, but also of other organisms. It turned out that despite the strong divergence of organisms over 2.5 billion years, ribosomal RNAs were quite conservative. Comparison of the corresponding elements was carried out with the help of programmes for mathematical processing of genome analysis.

Systematics. Phytophthora and malaria plasmodium

It was found that among bacteria, two groups (rRNA sequence tree) - Eobacteria and Archaebacteria - were distinguished on the basis of RNA similarity. It turned out that the Archaeobacteria group included such extreme habitat groups of prokaryotes as thermophiles and ultrathermophiles. Based on this, the first revolutionary step in systematics was made, which was brought by the study of DNA - among the kingdom of prokaryotes two subkingdoms were distinguished: Eobacteria and Archaea. It turned out that archaeal RNA has blocks that are not found in prokaryotes but are found in eukaryotes, i.e. their ribosomes are similar in some sense. They also have introns. Indeed, eukaryotes evolved from archaea.

Prokaryotes have few features for normal systematics. Bacteria have 3 (4) forms (vibrios and spirillae - curved twisted bacteria with flagella). In addition, contributed to the systematics of habitats, but this does not give the concept of a related system. Traits related to the assimilation of certain food sources (nutritional biochemistry) are still widely used. However, these traits are changed by single mutations. In the early 1960s, the parameter of DNA nucleotide composition was introduced for systematics. vertebrate rRNA contains few variations; in mammals, single substitutions of little significance occur. When comparing vertebrates with echinoderms, the difference can be seen: one can note how long certain substitutions took and indicate when the two groups diverged.

However, it turns out that this is difficult to trace on primitive organisms. There is an organism called Phytophthora, a parasite of solanaceous plants. When it attacks a tomato or potato, it takes 3-5 days from the onset of infection to the death of the entire plantation. When Phytophthora was tested for ribosomal RNA, it was found that it was not similar to what was thought to be related RNA of fungi. In fungi, the cell wall is made of chitin, while in phytophthora it is made of cellulose, like in plants. It turned out to be a relative of a green algae that had switched to parasitism by losing chlorophyll, so it started to look like a fungus.

Malaria plasmodium belongs to the sporozoites. Malaria plasmodium is a genospore - blood sporozoites. In them, a structure at the anterior end of the cell, the apicomplexa, was found in EM. This structure was also found in a number of other sporozoites, and soon this group was called apicomplexa. It turned out that their gene was closer to red algae, not to animals, as it was supposed (an alga that switched to parasitism). A chloroplasts turned into apicomplexa necessary for movement, metabolism, etc. Therefore, it is important to find a cure for malaria that inhibits plant metabolism.

If we build trees for eukaryotes, several levels emerge that are much higher than usually accepted (animal kingdom, plant kingdom, fungi kingdom). It turns out that these three kingdoms are the end of branching of a single branch that has branched many times. Separate autonomous systematic groups turned out to be euglenae. They are related trypanosomes (apparently, the first group that moved away from the common eukaryotic tree), infusoria (not related to either animals or fungi). This branching itself is quite complex, in which three levels are distinguished, preceding the notion of kingdom (clade). In some cases the division implies morphological differences. Infusoria and red algae and some other unicellulars have turned out to share a common structure that underlies the cell membrane on the outside (underneath it there are membrane vesicles that reinforce the cell surface (alveoli of membrane indentations) - this is the alveolate group. Llamas and camels, although ruminants, are not related to the parnopods, they are a separate branch (cerebellarians). Ruminants and non-ruminant parnopods are less related. Pinnipeds: long-eared and earless seals have different origins - one from predators, the other from ungulates.

Cetaceans appear to have evolved from an ancestor close to the hippopotamus, but it is not related to ungulates, although it has always been classified as a pair of ungulates. The general evolution of invertebrates is also different. Arthropods are related not to ringed worms but to roundworms. Consider the group of moulting and non-moulting invertebrates. Arthropod ringworms have to moult in order to grow. Their upper layers are tough but loosely bound to the epidermis, so they can shed them. This is due to the fact that the surface of epithelial cells in them is smooth, in other groups the surface of epithelial cells forms borders holding the cuticle, and they can not change it, because moulting in this case will lead to the rupture of cell membranes of the epidermis. So since they can't moult, they don't make such a hard cuticle, it's elastic and stretches. This was all done by analysing ribosomal RNA as the most conserved structures.

Structure of ribosomes

Each ribosome subunit contains one large RNA. When ionic conditions are created, the RNA folds into a compact structure (Fig. 13.1).

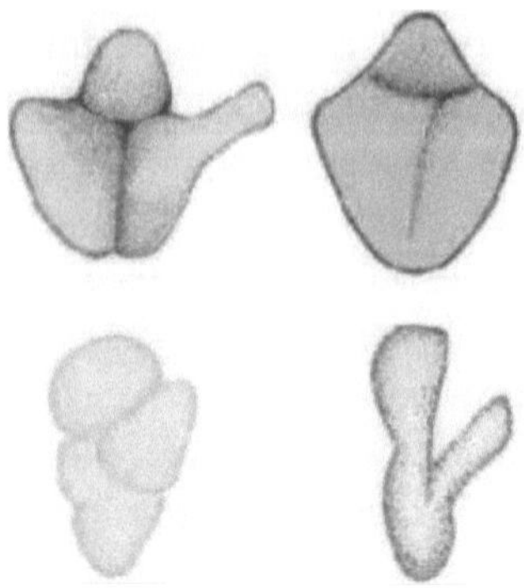

Figure 13.1: Types of ribosomal RNA. Top: 50S and 23S, bottom: 30S and 16S

The small subunit has a body, head and lateral lobe (top right in Fig. 13.1). The large subunit (top left in Fig. 13.1) also has a head, lateral lobe and a rather long extra finger. The darker part is the concave surface. These two subunits overlap, with the heads and body coinciding - the association-dissociation process of ribosomes is related to this. When the RNA is placed in the appropriate conditions, the subunits change their appearance (bottom figures respectively in Figure 13.1). In this case, the side lobe of the small subunit is deflected - there is a gap that is normally filled with proteins. In the large subunit, the finger (a purely protein structure) disappears, the head is reduced, and 5S RNA (transcribed in prokaryotes with others and excised) appears in the head. In eukaryotes, another 5.8S RNA turns up. If we compare the sequence of large ribosomal RNAs in prokaryotes and eukaryotes, it turns out that they do not have 5,8S RNA, but there is a corresponding sequence, which is included in 23S RNA. That is, the 5,8S is a cut off piece from the 28S at the point where the RNA has already folded and the ribosome has already started to assemble. Therefore, if RNA is isolated at room temperature without hard denaturation, 5,8S RNA and 28S RNA appear to be bound in a single complex, they can be dissociated only by breaking hydrogen bonds. That is, the shapes of the subunits in the upper figures differ from the lower ones by the presence of proteins.

(5,8S + 5S + 28S) 18S

Эукариот ***80S ⇆ 60S+40S***

Прокариот ***70S ⇆ 50S+30S***

5S+23S рРНК 16S рРНК

Eukaryote

Prokaryote

In viruses, DNA (or RNA) is in the middle and the protein forms the shell. Unlike the similar proteins in viruses, the ribosome has a huge number of different proteins. The proteins make the stacking of the ribosome more robust under a wide range of conditions. Only one protein in the ribosome is not present in a single copy, but as four identical subunits that form a complex together (L7L12 - because of its existence in two forms under different conditions: acetylated and non-acetylated) forming a finger, and the fifth subunit of the finger - its base - is another protein (L10).

The ribosome is where the components for protein synthesis are assembled. First of all, the mRNA binds to the small subunit. and anchors itself to it. The ribosome is also bound to tRNA, which brings nucleotides. There are two sites for this purpose: on the small subunit in the immediate vicinity of mRNA binding, due to this site tRNA is fixed next to mRNA, due to which further interaction of codon and anticodon takes place. In turn, in a large part of the ribosome there is a site that is recognised by tRNA and fixes the amino acid on it (A-site). The tRNA also retains the peptide on the ribosome because the peptide is bound to the tRNA by its C-end during synthesis (this site is the P-site) (Fig. 13.2).

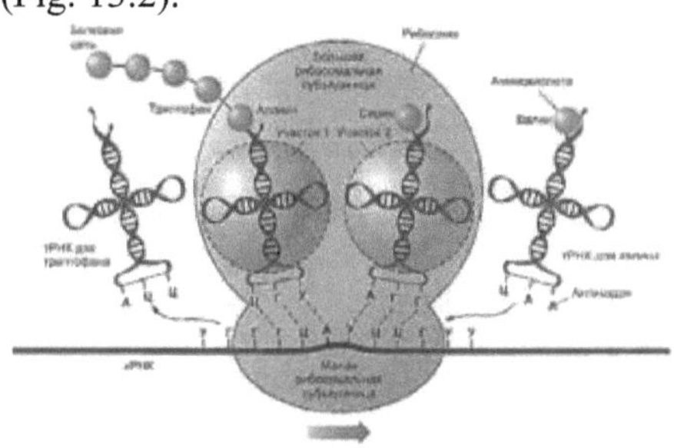

Fig. 13.2 Codon-anticodon interaction

In addition to the binding site on the ribosome there is a change site - the peptidyltransferase site (centre) - which catalyses the formation of a peptide bond. The ribosome also has a translation factor binding centre - located at the base of the finger.

Synthesis process. Initiation of translation

The synthesis process itself breaks down into three parts. First, the correct reading frame must be found. This is the initiation step. At this stage, ribosomes attach to mRNA at a strictly defined location to begin synthesis. The second stage is elongation. It consists of tRNAs that are complementary to the codon on the mRNA and carry their own amino acids, which helps to elongate the chain to form a peptide in the form of an alpha-helix. This is followed by a process called termination - initiated by the entry of stop codons into the ribosome, which completes the synthesis (Fig. 13.3).

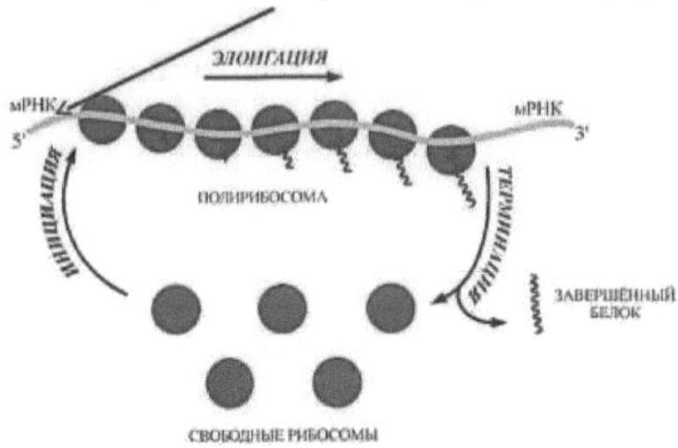

Fig. 13.3. Epicycle of translation

When assembled, the ribosome cannot start synthesis. Therefore, synthesis does not start with free ribosomes, but with free units. Synthesis starts with the AUG codon encoding the amino acid methionine, but you need to be able to distinguish the starting AUG codon from the one in the middle of the chain, for example. In eukaryotes there is a cap at the 5'-end, as soon as the mRNA enters the cytoplasm, special proteins (initiation factor IV, IF4) form a complex with this cap, to which the small subunit of the ribosome binds, recognising not the cap but the protein complex. Before the small subunit sits on the mRNA, it is joined by initiation factor II (IF2), which carries an initiator tRNA that helps the small subunit find the nearest AUG codon to the cap, from which it begins synthesis. The subunit is assisted in its movement by initiation factors 4a and 4b (RNA helicases). These factors in complex can bind ATP, bind to RNA and weave a helical structure on it, after which ATP hydrolysis, conformation change and detachment from RNA occurs. This continues until binding to the start codon occurs. At this point, due to the coupling of tRNA and mRNA, the ribosome is inhibited, and some proteins are also dissociated to link the small subunit of the ribosome to the large subunit. Next, elongation begins.

Prokaryotes have a difference. mRNAs are polycistronic, containing sequences encoding several proteins. This leads to the possibility of independent protein synthesis. This is not always the case, as proteins are encoded by a single mRNA and are needed together and in equal amounts. Sometimes there is an additional need for some proteins. In this case it is not possible to start the synthesis from the 5'-end.

Before each coding region there is a ribosome binding site for its independent operation. Such a site is in the region before the beginning of the coding sequence. It is six purine bases in a particular sequence: GGAGGA. It turns out that the complementary sequence to this one is the sequence in the small subunit RNA, the 3'-end of which is free. Thus, in prokaryotes, the small subunit attaches to the mRNA first, and the second subunit attaches to it. The initiating tRNA can join the small subunit earlier, or it can join the already prepared complex (Fig. 13.4).

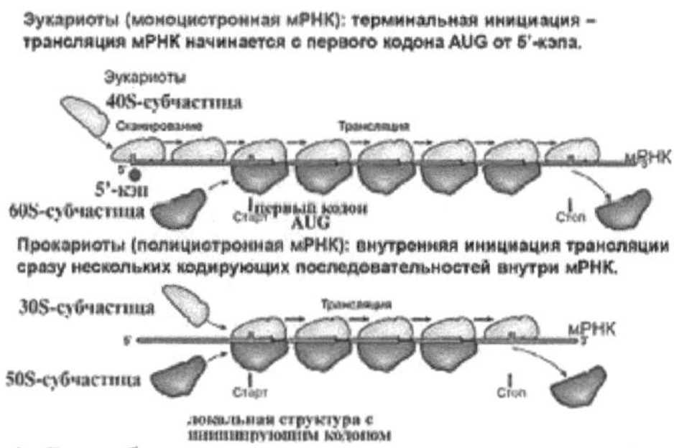

Figure 13. 4. Ways of recognising the initiator codon of mRNAu eukaryotes and prokaryotes

Another difference is that in prokaryotes the amino acid attached to the initiator tRNA is chemically modified. In eukaryotes it is phenylalanine, while in prokaryotes a formic acid residue (formylmethionine) is additionally attached to the amino group of phenylalanine to mimic the peptide. It is worth noting that this does not result in improved binding at the peptidyltransferase centre, as one might assume. When the ribosome is assembled, the factors involved in initiation begin to separate: first the third factor (IF3). The third factor, by attaching to the small subunit, causes the dissociation of the ribosome into small and large subunits. IF2, due to GTP hydrolysis, changes conformation and dissociates. Thus, the P-site remains occupied and the A-site is released. Next, the elongation step begins.

CHAPTER 14

14. Translation: elongation and termination

Initiation of translation

Let's finish the consideration of protein biosynthesis. The initial stage (initiation), one of the tasks of which is the correct setting of the reading frame. Reading starts with the AUG codon. After the codon is assembled, a complex consisting of large and small subunits of the initiator tRNA is assembled on it (Fig. 14.1).

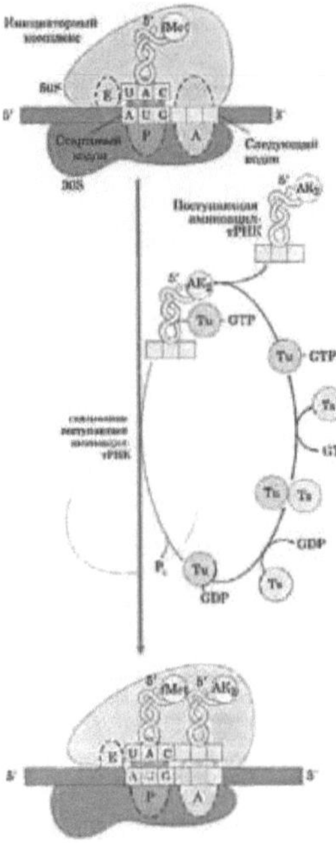

Fig. 14.1 Formation of initiator complex in bacteria

In the small subunit there is mRNA, which has an AUG codon in the p-partition and an initiator RNA is attached to it complementary to the anticodon, which carries methionine, and in bacteria it is additionally formed (a formic acid residue is attached to the KN2-group).

Elongation, cycle, factors, amino acid transfer. Translocation

Such a complex proceeds further to the next step, called elongation (elongation). Elongation occurs in accordance with the programme set by the mRNA. The A site in the ribosome is not 146
contains RNA, but contains mRNA, unlike initiation, any codons can end up here. If the next codon is UUC, a tRNA complementary to the anticodon will stand in the A-site of and bind to it (i.e. e. GAA). Such an

anticodon will be y tRNA, which recognised synthetase and on which the specific amino acid phenylalanine is attached. To speed up the process, an auxiliary protein called elongation factor comes into play. This protein is not part of the ribosome and functions independently and can attach to the ribosome. Just like the initiation factor two (IF2) that brought tRNA in the first step, this factor will bring tRNA and interact with the same site on the ribosomal subunit. That is, there is a binding factor located at the base of the thumb of the large subunit, and at that site the elongation factor attaches from the outside so that the tRNA that attaches to it faces inward. That is, the interaction is ensured energetically by the recognition of the factor and the ribosome, and the tRNA is inserted by this factor into the middle part. This is a TU (thermo unstable) factor, and it is present in two states. The first is bound to GTP and the second is bound to GDF, these are two different conformations. When bound to GTP, it can bind to tRNA to form a compact globule. When bound to GDF, it has a domain divergence that makes it more loose and the binding centre to the ribosome is stretched in space. The same happens with the tRNA binding centre. That is, the factor in complex with GTP (TU-GTP) interacts with aminoacyl-tRNA, forming a complex that sits in the A-part of the ribosome. The anticodon sits complementarily, with the amino acid inside on the large subunit (the A-unit captures the two subunits and the space between them). This is some catalysis, but not chemical catalysis, but binding catalysis, essentially. The 3'-end on which the tRNA sits appears to bind to the factor, and once in the ribosome, the tRNA binds firmly and correctly to the anticodon, while its amino acid end is blocked. Therefore, it is necessary to remove the factor to bind to the peptidyl transferase centre. The removal occurs by hydrolysis of GTP, and the factor itself contains an active centre that hydrolyses GTP. Due to the loss of phosphate, domains 1 is stretched from 2 and 3, causing the affinity for the ribosome and tRNA to be lost and the factor to leave (Fig. 14.2).

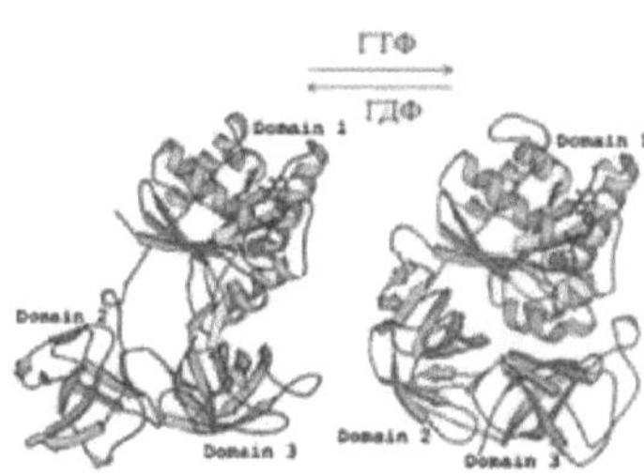

Figure 14.2: Elongation factor (left GDF-form (TU-GDF), right GTP (TU-GTP))

Thus, the amino acid end enters the peptidyltransferase centre. This process can proceed without the factor, the process follows the same principle, but at a much slower rate. Thus, the p-site is occupied by the initiator tRNA, and the A-site is occupied by the next tRNA, and the amino acids attached to these two tRNAs are brought together in the peptidyltransferase centre. It is worth noting that TU-GDF is quite stable, but continuous operation of the factor is required. In order to remove GDF, a second protein is needed which, by interacting with the factor, will displace GDF. This is TS (thermo stable). TS displaces GDF by binding to TU to form a complex. This in turn leads to the possibility of TU binding to GTP, allowing the factor to fulfil its role again.

When two amino acids are attached to a tRNA, the peptidyltransferase centre comes into play (Fig. 14.3).

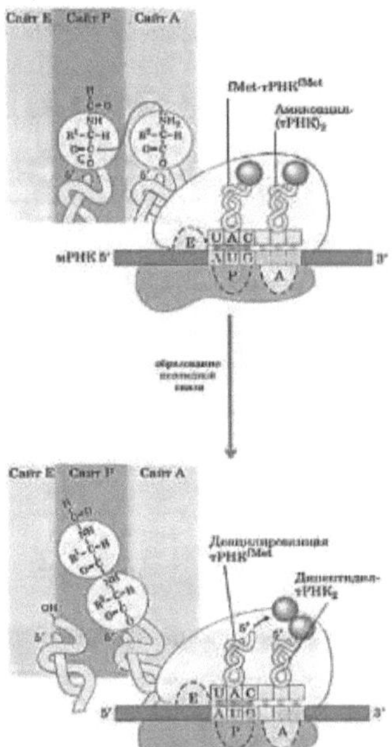

Figure 14.3: The second stage of elongation in bacteria: formation of the first peptide bond

The centre carries out the chemical transfer of the amino acid from the p-site from the tRNA to the amino group of the amino acid from the A-site. Once the amino group of the A-amino acid attacks the keto group of the p-amino acid to form a bond, the end of the tRNA is released, which the ribosome pulls back to the E-site, which binds the 3'-end of the tRNA. The resulting dipeptidyl-tRNA is located between the A- and P-partitions.

Translocation takes place in mRNA to read the next codon. It also takes place with the help of a special protein factor (G-factor, GTP-dependent factor). Being a protein, this factor partially mimics tRNA (molecular mimicry). G-factor binds to GTP or its analogue. In addition to domains 1-3, G-factor also has domains 4 and 5 that mimic tRNA: the alpha-helix contains many negatively charged amino acid residues. The molecule binds like TU to the ribosome at the factor-binding centre, with the tRNA-mimicking part going to the A site. The tRNA bound to the dipeptide from the A-site is displaced deep into the ribosome - into the p-site, and the tRNA from the p-site is shifted to the E-site, thus the mRNA is shifted by 3 nucleotides. The first (initiator) tRNA enters the E-partition and passes through it to the outside, the second tRNA with the dipeptide ends up in the P-partition, the A-partition remains occupied by the factor, which, like TU, hydrolyses its ATP, which leads to loosening and degradation of the factor, and the A-partition is released. At this point the elongation cycle ends and everything repeats. Each cycle the peptide is lengthened by one amino acid, the matrix is shifted by 3 nucleotides, and the process proceeds according to the genetic text written in the mRNA. (Figure 14.4).

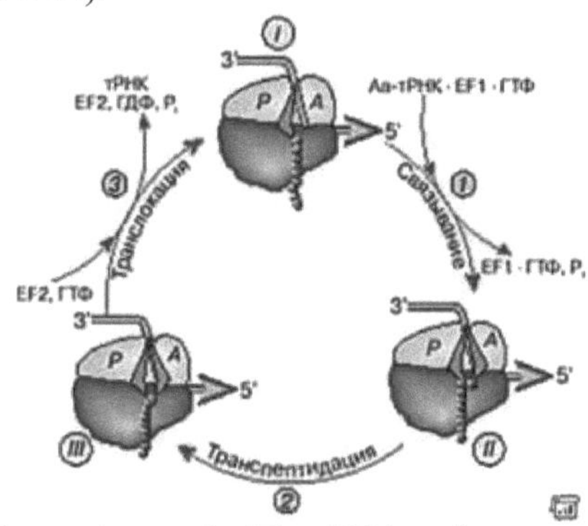

Figure 14.4: The elongation cycle. The tRNA is shown in green and the peptide coupled to it in blue. EF1 and EF2 are elongation factors (in the case of bacteria, TU and G-factor, respectively)

Binding of the factor to the ribosome is non-specific and occurs at any tRNA, hence errors occur: at the stage of interaction with TU and incorporation into the ribosome - 1% each (1 error per 10 thousand cases).

Antibiotics that affect ribosomes

Since this selection is the most complex process, the bottleneck, it is easy to attack it with specific antibiotics. The class of antibiotics aminoglycosides are derivatives of carbohydrates in which several hydroxyls are replaced by amino groups. The amino groups are protonated and positively charged and, once in the ribosome (low molecular weight compounds, easily diffuse inside), bind to RNA, forming cationic bridges between tRNA and rRNA in the A-partition. Consequently, the tRNA gets stuck in the ribosome, and the one with the wrong codon-anticodon interactions is the one that should have dissociated. These are bactericidal antibiotics, causing bacterial death if they are specific to bacterial ribosomes. 50% of all antibiotics target the ribosome (streptomycin, see lecture 2). Other antibiotics affect other steps in this cycle (fusidic acid acts on translocation, stopping cell growth). Tetracyclines (can be extensively chemically modified) act on tRNA binding by preventing the end of the tRNA carrying an amino acid from binding properly, taking the place of that end, and preventing the tRNA from lingering in the ribosome. Levomecitin (chloramphenicol) blocks the peptidyltransferase centre and does not block the trans-peptidation reaction. Erythromycin (macrolide antibiotics) binds near the peptidyltransferase centre, does not interfere with the formation of the dipeptide bond, but does interfere with the exit of the resulting peptide from the ribosome, thus stopping synthesis.

Termination

The elongation cycle is repeated until the ribosome reaches the terminating codon. The terminating codon does not have a tRNA. Additional protein molecules (termination factors) are required to carry out the termination process. Termination occurs in response to a stop codon in the A site. First, the termination factor RF (RF1-1 or RF-2, depending on the stop codon) binds to the A site. This leads to hydrolysis of the ester bond between the polypeptide and tRNA at the P site and release of the terminated polypeptide. Finally, the mRNA, deacylated tRNA, and termination factor leave the ribosome, which dissociates into 30S- and 50S-subunits with the participation of ribosome recycling factor (RRF), initiation factor IF- 3, and by energy released by GTP hydrolysis mediated by factor EF-G. The complex of the 30S subunit with IF-3 is ready to start the next translation cycle (Fig. 14.5).

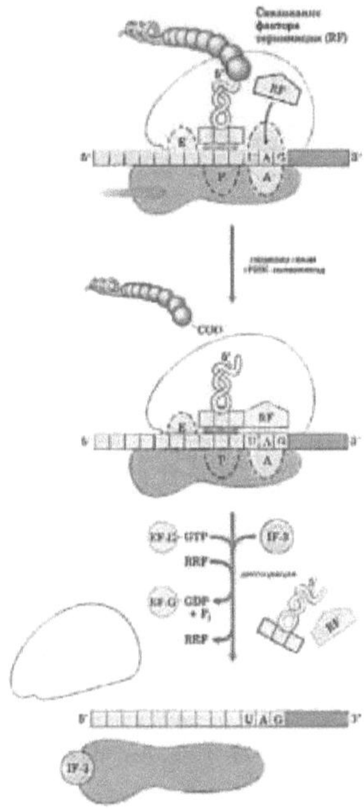

Fig. 14.5. Translation termination

In eukaryotes, the process is slightly different, but follows the same scheme, with the only difference being that the further fate of the ribosome is not completely clear. But it happens that after reading the gene, the ribosome, freed from tRNA, does not dissociate, but begins to slide along the mRNA, which leads to a shift in the place of the end of synthesis. In this case, it can bump into the Shain-Dalgarno (Sh-D) sequence in bacteria. This is especially true for prokaryotes because they have polycistronic matrices, meaning there are multiple genes on the mRNA and they have to work in a coordinated fashion because they are working in the same processes. If there is a Sh-D sequence, the small subunit clings to it and then the initiator tRNA is attached.

Eukaryotes also have a phenomenon called head and tail dialogue. There are specific proteins that bind cap and a specific poly-A-binding protein. Due to the fact that poly-A is long, it forms a structure similar to chromatin: a densely packed area where this protein is bound to the adenyl sequence. It turned out that some cap proteins interact strongly with poly-A proteins, which leads to RNA cyclisation and accelerated initiation.

Transport of synthesised proteins

It is worth paying attention to the transport of synthesised proteins in eukaryotes. This is the rough endoplasmic reticulum. The structure permeates the whole cell, it is rough because it is covered by ribosomes and synthesises a protein that must have a signalling (start) sequence at the N-terminus. The SRP particle is involved in this process, is a 7S-PHK complex with proteins that recognises the signal peptide and binds to it. When the ribosome starts to synthesise RNA, it lands on the initiating codon, passes it and starts to synthesise the protein. Once the signal sequence has gone beyond the ribosome, it binds the SRP particle, but the SRP binds not only to this end but also to the ribosome, plugging it, forming a complex that cannot synthesise the protein any further. This complex ends up on the membrane of the endoplasmic reticulum. This membrane has a special receptor that recognises and binds SRP, and it also interacts with membrane proteins of the same type but which form a kind of oligomer, forming a channel whose components firmly bind the ribosome and allow the synthesised peptide to pass through and enter the endoplasmic reticulum. A signal peptidase binds to the peptide and cuts off the signal sequence (Fig. 14. 6).

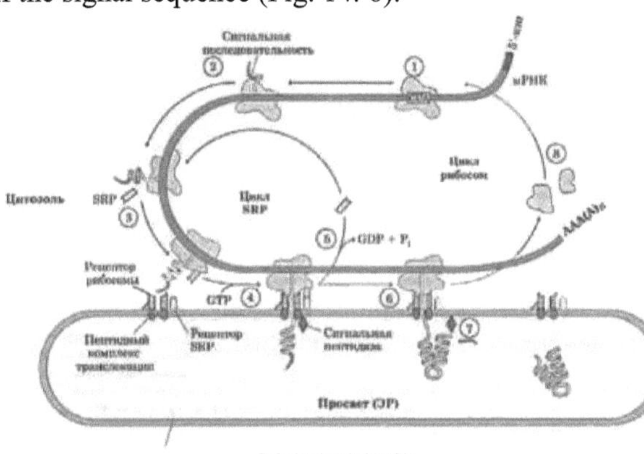

Figure 14.6: Delivery of eukaryotic proteins with signalling proteins sequences into the endoplasmic reticulum.

Glycosylation

The protein then undergoes glycosylation. For example, by serine or threonine (o-glycosylation). A specific site is glycosylated through the attachment of single monosaccharides. N-glycosylation - attachment by the amide group of asparagine, and already ready oligosaccharide, previously synthesised. These processes also occur as the protein is synthesised (Fig. 14.7).

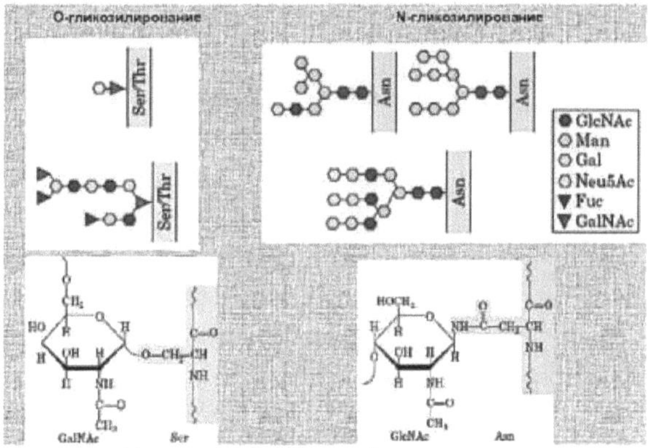

Figure 14.7 Glycosylation options

Problem solving in molecular biology

The student should know

1) the structure of DNA and RNA;
2) basic molecular biology terminology necessary to understand and solve typical molecular biology problems;
3) the rules of complementarity of base pairs;
4) Chargaff rules;
5) types of RNA and their classification;
6) the average mass of nucleic acids and amino acids;
7) the distance between nucleotides in the DNA chain;
8) a three-nucleotide code;
9) methods of molecular biology (session 10).

The student should be able to

1) solve typical problems in molecular biology;
2) explain the process of solving a typical problem;
3) use the codon table.

Molecular biology studies the mechanisms of storage and transmission of hereditary information. Molecular biology tasks are found in two main topics:
nucleic acids, the genetic code.

Basic knowledge required to solve a molecular biology problem The genetic code is a system for recording genetic information about the sequence of amino acids in proteins as a sequence of nucleotides in DNA or RNA.

Properties of the genetic code:

1. The code is triplet. One amino acid is encoded by three nucleotides.
2. The code is universal. All living organisms (from bacteria to humans) use the same genetic code.

3. The code is degenerate. One amino acid is encoded by more than one triplet.
4. The code is unambiguous. Each triplet corresponds to only one amino acid.
5. The codon does not overlap. One nucleotide cannot be part of several codons in the mRNA chain.

The sequence of nucleotides in a DNA molecule determines its specificity, as well as the specificity of the organism's proteins that are encoded by that sequence. These sequences are individual to each species of organism and to the individuals of a species.

DNA (deoxyribonucleic acid)two chains in a helix RNA (ribonucleic acid) one chain consist of nucleotides Structure of a nucleotide

1 deoxyribose
2 phosphoric acid residue
3 nitrogenous base

A - adenine
G - guanine
C - cytosine
T - thymine

ribose 1
phosphoric acid residue 2
nitrogenous base 3

A - adenine
G - guanine
CD - cytosine
U - uracil

Principle of complementarity of bases A-T, G-C A-U, G-C Complementarity is due to the number of hydrogen bonds A = T double, G = Ts triple DNA RNA

Chargaff's rule

1. A=T, D=TS ; A+G = T+TS ^ 100%
50% 50%
(in two circuits)
A + D + Y + TS = 100%
(in the same circuit)

Nitrogenous bases: 1. Purine bases - A, G 2. Pyrimidine bases - Ts, T, U

Chargaff's Rule: Σ(A)= Σ(T), Σ(Г) = X(Ts), Σ(A+Г) =E(T+D) One step is a complete turn of the DNA helix - a 360o turn One step is 10 pairs of nucleotides

The length of one step is 3.4 nm
The distance between two nucleotides is 0.34 nm
The molecular mass of one nucleotide is 345 g/mol
Molecular weight of one amino acid - 120 g/mol
The average molecular weight of one amino acid residue is taken as 120.
The relative molecular mass of one nucleotide is taken as 345. Calculation of the molecular mass of a protein: Mmin = (a : c) x 100%, where: Mmin is the minimum molecular mass of the protein, "a" is the atomic or molecular mass of the component, "c" is the percentage content of the component.

Types of tasks

All tasks in molecular biology can be categorised into seven types:

1. Establishing the sequence of nucleotides in DNA, iRNA, tRNA anticodons using the principle of complementarity.
2. Calculation of the number of nucleotides, their percentage in the DNA chain, iRNA.
3. Calculation of the number of hydrogen bonds in the DNA, iRNA chain.
4. Determination of length, mass of DNA, iRNA, gene.
5. Determining the sequence of amino acids from the genetic code table.
6. Determination of the amount of amino acids, nucleotides.
7. Combined tasks.

Requirements for problem solving

□ the course of the solution must correspond to the sequence of processes occurring in the cell;
□ solve problems consciously, justify each action theoretically;
□ Record the solution neatly: DNA, iRNA, tRNA chains are straight, nucleotide symbols are clear, located on one line horizontally;
□ DNA, iRNA, tRNA chains to be placed on one line without hyphenation;
□ write the answers to all questions at the end of the solution.

The first type of tasks are tasks to establish the sequence of nucleotides in DNA, iRNA, tRNA anticodons

A section of the right chain of the DNA molecule has the nucleotide sequence: A-G-T-T-C-T-A-A-C-T-G- A-G-C-T-A-T. Write down the nucleotide sequence of the left DNA chain.

Given: DNA A-G-T-T-T-T-A-A-T-T-T-G-A-G-T-T-A-T.

Solution: (we select the nucleotides of the left DNA chain according to the principle of complementarity A-T, G-C) DNA A G T T T A T A T A T A T A T G A T A T A T A T A T A T A T A T A T A T A T A T A T A T A T A T A T A T A T A T A T A T A T A T A T A T A T A T A T A T A

Answer: the left strand of DNA has a nucleotide sequence of T-C-A-G-A- T-

T-G-A-T-T-T-T-T-T-C-G-T-A

The second type of tasks for calculating the number of nucleotides, their percentage ratio in the DNA chain, iRNA

One DNA molecule contains 22% of nucleotides with thymine T. Determine the percentage of nucleotides with A, G, CD individually in this DNA molecule.

Given: T -22%

Find: % A, D, C

Solution 1.

1) According to Chargaff's rule A+G = T+C, all nucleotides in DNA are 100%. Since thymine is complementary to adenine, A=22%. 22+22=44% (A+T) 100- 44 =56% (G+TS)

2) Since guanine is complementary to cytosine, their numbers are also equal, therefore

56 : 2 =28% (G, TS)

Solution 2:

1) According to Chargaff's rule A+G= T+C, all nucleotides in DNA are 100% or A+G and T+C are 50% each As thymine is complementary to adenine, A=22%. hence 50 - 22=28% (G, C as they are complementary).

Answer : A=22%, D=28%, C=28%

How many nucleotides A, T, G, are contained in a fragment of a DNA molecule if it contains 1500 nucleotides of CD, which is 30% of the total number of nucleotides in this DNA fragment?

Given: Ts - 30% = 1500 nucleotides

Find: number of nucleotides A, T, G

Solution:

Answer: a fragment of DNA molecule contains: G=1500 nucleotides, A=1000 nucleotides, T=1000 nucleotides.

A section of a DNA molecule (one strand) contains: 150 nucleotides - A, 50 nucleotides - T, 300 nucleotides - CD, 100 nucleotides - G.

Determine: the number of nucleotides in the second chain with A, T, G, CD and the total number of nucleotides with A, T, CD, G in the two DNA strands.

Given: nucleotides in the 1st strand of DNA: A - 150, T - 50, Ts - 300, G - 100.

Find: A, T, C, G in two strands of DNA.

Solution:

Answer: nucleotides in the second chain T - 150, A - 50, G - 300, CD - 100; 1200 nucleotides in two chains.

Using the genetic code table, complete the table below. Mark the 5'- and 3'- ends of the nucleic acids and the N- and C- ends of the polypeptide. Transcription and translation proceed in a left-to-right direction.

C												2x цепочечная ДНК
						T	G	A				
	C	A				U						мРНК
									G	C	A	Антикодоны тРНК
				Trp								Аминокислоты, составляющие белок

Решение:

C	G	T	A	C	C	A	C	T	G	C	A	2x цепочечная ДНК
G	C	A	T	G	G	T	G	A	C	G	T	
G	C	A	U	G	G	U	G	A	C	G	U	мРНК
C	G	U	A	C	C	A	C	U	G	C	A	Антикодоны тРНК
	Ala			Trp			-			Arg		Аминокислоты, составляющие белок

The third type of problems for calculating the number of hydrogen bonds.Two DNA chains are held together by hydrogen bonds. Determine the number of double and triple hydrogen bonds in this DNA chain if it is known that there are 12 nucleotides with adenine and 20 with guanine in both chains.

Given: A - 12, D - 20

Find: the number of 2 and 3, hydrogen bonds in DNA

Solution:

Answer: 24 double hydrogen bonds, 60 triple hydrogen bonds, for a total of 84 hydrogen bonds.

The fourth type of problems is length determination, DNA, iRNA.A section of of the DNA molecule consists of 60 nucleotide pairs. Determine the length of this section (the distance between nucleotides in DNA is 0.34 nm)

Given: 60 pairs of nucleotides

Find: the length of the section

Solution: nucleotide length 0.34 nm 60*0.34= 20.4 nm

Answer: 20.4 nm

The number of nucleotides in the DNA chain is 100. Determine the length of this section.

Given: 100 nucleotides

Find: the length of the section

Solution: nucleotide length is 0.34 nm , DNA consists of 2 chains so 50 pairs of nucleotides. 50*0.34=17nm

Answer: 17nm

The number of nucleotides in the iRNA chain is 100. Determine the length of this section

Given: 100 nucleotides

Find: the length of the section

Solution: nucleotide length is 0.34 nm , iRNA consists of one chain 100x0.34=34nm

Answer: 34nm

A section of one DNA strand contains 720 nucleotides, 120 of which are introns. Determine the length of the original pre-iRNA, the length of the mRNA and the number of amino acids included in the synthesised polypeptide.

Solution:

1. We determine the number of nucleotides in the composition of iRNA, knowing that it is equal to the number of DNA exons.

720 - 120 = 600

2. We determine the length of pre-iRNA and iRNA, knowing that the length of one nucleotide is 0.34 nm. length of pre-iRNA: 0.34 x 720 nucleotides = 244.8 nm length of iRNA: 0.34 x 600 nucleotides = 204 nm

3. We determine the number of amino acids in the polypeptide, knowing that one amino acid is encoded by three nucleotides.

600 : 3=200 amino acids.

Studies have shown that iRNA contains 34% guanine, 18% uracil, 28% cytosine, and 20% adenine. Determine the percentage composition of the nitrogenous bases in the section of DNA that is the matrix for this iRNA.

Solution (for convenience, we use the tabular form of recording the solution):

[-RNA	GUNA			
	34%	18%	28%	20%
DNA (semantic strand,		A	ц	
readable)	28%	18%	34%	20%
DNA (antisense strand)	Г	A	ц	T
	34%	20%	28%	18%

We calculate the percentage of nitrogenous bases based on the principle of complementarity:

The sum of A+T and G+C in the semantic chain will be:

A+T=18%+20%=38%; G+TS=28%+34%=62%. In the antisense chain (non-coding) chain, the totals will be the same, but the percentage of individual bases will be reversed: A+T=20%+18%=38%; D+C=34%+28%=62%. In both chains the pairs of complementary bases will be equal, i.e., adenine and thymine - 19% each, guanine and cytosine - 31% each.

The molecular mass of the DNA fragment is 90,000. Of the total number of nucleotides in this fragment, 85 are thymine nucleotides. Determine the number of guanine, cytosine, and adenine nucleotides in this DNA fragment. What is the length of this DNA fragment?

Solution:

1. Determine the total number of nucleotides in the two DNA chains, knowing that the average molecular weight of one nucleotide is 345.

90,000 : 345 = 260 nucleotides

2. Application of Chargaff's rule. There are 85 thymine nucleotides in the fragment, hence 85 adenine nucleotides. Find the number of cytosine and guanine nucleotides.

260 - (85 + 85) = 90

90 : 2 = 45 (G or Ts).

3. We determine the length of the DNA fragment, knowing that it contains 260 nucleotides in total, 130 in each strand.DNA length = 0.34 nm *130 = 44.2 nm.

"Protein biosynthesis, the genetic code."

- □ pre-iRNA is built on a section of DNA
- □ iRNA is formed as a result of processing and splicing
- □ iRNA moves into the cytoplasm
- □ iRNA binds to the ribosome (2 triplets)
- □ tRNA carries an amino acid into the ribosome
- □ the iRNA codon is complementary to the tRNA anticodon
- □ protein is formed from amino acids in the ribosome
- □ DNA - iRNA-tRNK
- □ DNA - iRNA protein
- □ 20 amino acids - 61 triplets
- □ 3 nucleotides =1 triplet =1 amino acid = 1tRNA

To the question about the number of tRNA types, in textbooks one can find a typical reasoning that there can be 64 possible tRNA types (the number corresponds to the number of combinations of codons and 4 to 3) minus three stop codons, and the total is 61.

But, besides canonical amino acids, there are two non-canonical ones -

selenium cysteine and selenium methionine, which are included in the synthesised peptide when reading stop codons that are in a special nucleotide environment, meaning that it is not a stop codon, but a codon of a selenium-containing amino acid. Accordingly, there must be more than 61 tRNAs.

But, mitochondria of almost all species contain a minimalistic set of tRNAs - 22 species (Suzuki T, Nagao A, Suzuki T. Human mitochondrial tRNAs: biogenesis, function, structural aspects, and diseases. Annu Rev. Genet. 2011;45:299-329). Whereas, in the cytoplasm, the number of tRNAs varies greatly from species to species and usually occurs around 50 types for 22 amino acids. However, the same set of tRNA types does not usually occur in different species.

Towards the number of tRNA types.But, tRNAs undergo more than 100 types of modifications, effectively increasing the size of the nucleotide alphabet used by tRNAs (Machnicka MA, Milanowska K, Oglou OO, Purta E, Kurkowska M, et al. 2013. MODOMICS: a database of RNA modification pathways-2013 update. Nucleic Acids Res. 41(database issue:) D262-67).

However, the number of tRNA molecules formed in the cell depends on the number of copies of tRNA genes in the genome. Thus, there are 417 functional tRNA genes in the human genome (Chan P.P., Lowe T.M. 2016. GtRNAdb 2.0: an expanded database of transfer RNA genes identified in complete and draft genomes. Nucleic Acids Res. 44(database issue:) D184-89). The most sought-after tRNAs are expressed. Do tRNAs make errors in recognising amino acids - they do, the frequency of translation errors ranges between 10-4 and 10-3 per amino acid.

To the question of the number of tRNA types. So, how many types of tRNAs exist in the cytoplasm?

A minimum of 31 tRNAs is required for unambiguous translation of all 61 sense codons; the maximum observed number is 41 (Lodish H, Berk A, Matsudaira P, Kaiser CA, Krieger M, Scott MP, Zipursky SL, Darnell J. (2004). Molecular Biology of the Cell. WH Freeman: New York. 5th ed.). But among the zoo of tRNAs are 16 more ambiguous ones that can recognise similar amino acids.

The fifth type of tasks is determining the sequence of amino acids using the genetic code table.

A fragment of a DNA chain has the nucleotide sequence: TGGGGAGTGGAGTTA. Determine the nucleotide sequence on the iRNA, the anticodons of the tRNA, and the amino acid sequence of a fragment of a protein molecule. Given: DNA T-G-G-A-G-A-G-T-G-A-G-G-A-G-T-T-A Find: iRNA, tRNA, and amino acid sequence of a protein.

Solution:

1) construct iRNA on a DNA strand according to the principle of complementarity (A-U, G-C), then construct tRNA on the iRNA chain according to the principle of complementarity (A-U, G-C).

DNC T- G- G- A-G- T- G- A- G- T- T- A

iRNA A-C-C-C-U-C-A-C-C-U-C-A-C-U-C-A-A-A-A-U

tRNA U- G-G-A-G-A-G-G-U-G -A- G-G-U-U-U-A

2) divide the iRNA into triplets and determine the amino acid sequence of the protein using the genetic code table: A-C-C-Tre, U-C-A Ser, C-U-C Leu, A-A-U Asn.

Answer: iRNA A-C-C-C-U-C-A-C-U-C-A-C-U-A-C-U-A-U-A-U

tRNA U-G-G-A-G-U-G-U-G-A-G-G-U-U-U-A amino acid sequence of protein :tre, ser, leu, asn

Codon table

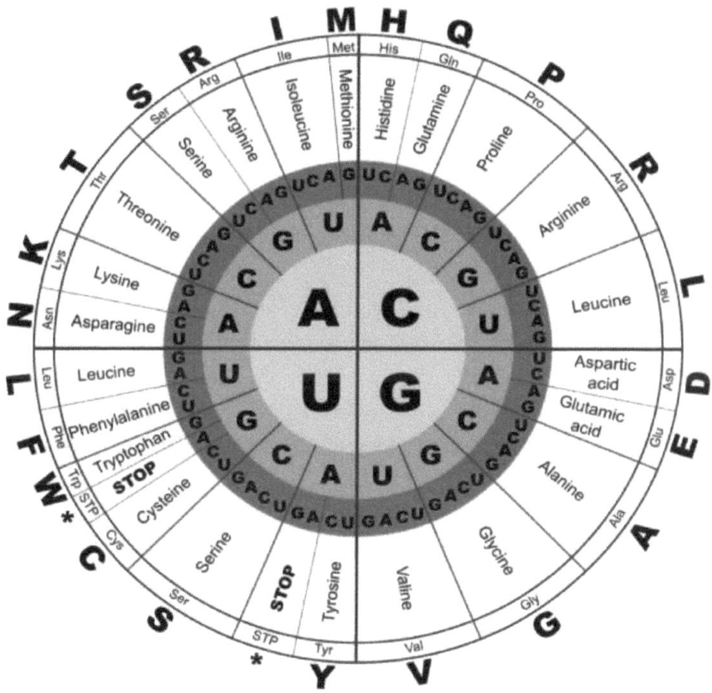

Второй нуклеотид

Первый нуклеотид	У	Ц	А	Г	Третий нуклеотид
У	УУУ, УУЦ – Фенилаланин; УУА, УУГ – Лейцин	УЦУ, УЦЦ, УЦА, УЦГ – Серин	УАУ, УАЦ – Тирозин; УАА Стоп-кодон; УАГ Стоп-кодон	УГУ, УГЦ – Цистеин; УГА Стоп-кодон; УГГ Триптофан	У Ц А Г
Ц	ЦУУ, ЦУЦ, ЦУА, ЦУГ – Лейцин	ЦЦУ, ЦЦЦ, ЦЦА, ЦЦГ – Пролин	ЦАУ, ЦАЦ – Гистидин; ЦАА, ЦАГ – Глутамин	ЦГУ, ЦГЦ, ЦГА, ЦГГ – Аргинин	У Ц А Г
А	АУУ, АУЦ, АУА – Изолейцин; АУГ Метионин старт-кодон	АЦУ, АЦЦ, АЦА, АЦГ – Треонин	ААУ, ААЦ – Аспарагин; ААА, ААГ – Лизин	АГУ, АГЦ – Серин; АГА, АГГ – Аргинин	У Ц А Г
Г	ГУУ, ГУЦ, ГУА, ГУГ – Валин	ГЦУ, ГЦЦ, ГЦА, ГЦГ – Аланин	ГАУ, ГАЦ – Аспарагиновая кислота; ГАА, ГАГ – Глутаминовая кислота	ГГУ, ГГЦ, ГГА, ГГГ – Глицин	У Ц А Г

A section of the DNA molecule has the following structure: GGA-ACC-ATA- GTC-CAA

Determine the nucleotide sequence of the corresponding section of iRNA. Determine the sequence of amino acids in the polypeptide synthesised by iRNA. How will the sequence of amino acids in the polypeptide change if the fifth nucleotide in DNA is replaced by adenine as a result of the mutation? Explain your answer.

Given: DNA GGA -ACC-ATA-GTC-CAA

Find: amino acid sequence of the original, mutated protein

Solution:

Answer: pro, tri, tir, gln, val; pro, ley, tir, gln, val.

It is known that all types of RNA are synthesised on a DNA matrix. The fragment of the DNA molecule on which the central loop of tRNA is synthesised has the following nucleotide sequence ATAGCTCTGAATCGGGATCT. Determine the nucleotide sequence of the section of tRNA that is synthesised on this fragment and the amino acid that this tRNA will carry during protein biosynthesis if the third triplet corresponds to the anticodon of the tRNA.

Given: DNA : ATAGCTCTGAAGCGCGCGCT

Find: the nucleotide sequence of a section of tRNA the amino acid that the tRNA will carry

Solution:

1)Since tRNAs are synthesised on DNA, let us construct tRNAs according to the principle of complementarity (A-U, G-C) DNA A T A G C T G A C G A C G A C T tRNA U A U C G A C G C U G C G C C A

2) An amino acid is encoded by an iRNA codon . tRNA U A U U C G A C U U G C U G C C C U G A

iRNA G A A A

The third triplet (anticodon of tRNA) TSUU , corresponds to the codon on iRNA GAA (according to the principle of complementarity), according to the table of genetic code this codon corresponds to the amino acid GLU, which is carried by this tRNA.

The answer is: tRNA UAUTSGAUTSUGCUUGCUGA amino acid GLU.

In Fanconi syndrome, the patient excretes amino acids with urine, which correspond to codons in iRNA: AUA GUG AUG UCA UCA UUG UUG GUG UUU AUU. Determine which amino acids are excreted in the urine in Fanconi syndrome, if a healthy person's urine contains the amino acids alanine, serine, glutamic acid, and glycine.

Solution (for convenience we use the table form of recording the solution): Using the principle of complementarity and the table of genetic of the code we get:

иРНК	АУА	ГУЦ	АУГ	УЦА	УУГ	ГУУ	АУУ
Аминокислоты цепи белка (больного человека)	Изе-Вал-Мет-[illegible]-Лей-Вал-Иле						
Аминокислоты цепи белка (здорового человека)	Ала-Сер-Глу-Гли						

Thus, in the urine of a sick person only one amino acid (serine) is the same as in a healthy person, the rest are new, and the three characteristic of a healthy person are absent.

The sixth type of problems - determination of protein mass, number of amino acids, nucleotides A fragment of a DNA molecule contains 1230 nucleotide residues. How many amino acids will be included in the protein?

Given: 1230 nucleotides.

Find: the number of amino acids

Solution:

One amino acid corresponds to 3 nucleotides, so 1230:3= 410 amino acids.

The answer is 410 amino acids.

How many nucleotides does a gene encoding a protein of 210 contain? amino acids?

Determine the number of amino acids that make up the protein, the number of triplets, and the number of nucleotides in the gene that encodes this protein if 30 tRNA molecules were involved in the translation process.

Given: 30tRNA

Find: number of amino acids, triplets, nucleotides in the gene

Solution:

Answer: amino acids 30, triplets 30, 90 nucleotides.The molecular mass of a polypeptide is 40000. Determine the length of the gene encoding it if the molecular mass of one amino acid is 100 on average and the distance between neighbouring nucleotides in the DNA chain is 0, 34 nm.

Given: mass of protein - 40000

amino acid mass -100

nucleotide spacing 0.34nm

Find: the length of the gene

Solution:

Answer: gene length 408 nm

The DNA of a phage has a molecular mass of 106. How many proteins can be encoded in it, assuming that a typical protein consists of 50 amino acids? Determine whether the gene or the protein has a larger molecular mass.

Solution:

1.Determine the molecular weight of one protein knowing that the average molecular weight of an amino acid is 120. 50*120=6000

2. determine the molecular mass of one gene. According to the problem condition, a typical protein consists of 50 amino acids, so one gene consists of 50 triplets (150 nucleotides, respectively).

It is known that the average molecular mass of a nucleotide is 345 150*345=51750

3. determine the number of proteins encoded in the phage DNA.

106 = 1 000 000

1 000 000 : 51750 = 19

Thus the molecular weight of the gene is greater than the molecular weight of the protein.

Combined tasks

A protein consists of 100 amino acids. Determine by how many times the molecular mass of the gene coding for this protein exceeds the molecular mass of the protein, if the average molecular mass of an amino acid is 110 and the average molecular mass of a nucleotide is 300.

Given:

100 amino acids, the molecular mass of an amino acid is 110, the molecular mass of a nucleotide is 300.

Find: how many times the mass of the gene is greater than the mass of the protein.

Solution:

1) Since a gene is a section of DNA consisting of nucleotides, let's determine their number: amino acids are 100, one amino acid is encoded by 3

nucleotides, then 100*3=300 nucleotides.

2) A protein is made up of amino acids. The molecular mass of protein is 100*110=11000, 3) A gene consists of nucleotides. The molecular mass of a gene is 300*300=90000

4) The molecular mass of the region of the gene encoding this protein exceeds the molecular mass of the protein: 90000 : 11000=8 times

Answer: 8 times

How long is the section of DNA molecule in which the primary structure of insulin is encoded if the insulin molecule contains 51 amino acids and one nucleotide occupies 0.34 nm in the DNA strand? What is the number of tRNA molecules needed to transport this number of amino acids to the site of synthesis? (Note that one tRNA delivers one amino acid to the ribosome.) Explain your answer.

Answer: DNA length 52 nm, 51 molecules of tRNA

Energy exchange

Basic concepts

1. Preparatory step (in the alimentary canal, lysosomes) starch glucose (E)
2. Oxygen-free "glycolysis" (in cytoplasm) glucose 2 PBC + 2ATP
3. Oxygen "respiration" (in mitochondria) PBC CO2 +H 2O + 36 ATP 1 glucose = 38 ATP

During glycolysis, 42 molecules of pyruvic acid were formed. How many glucose molecules were cleaved and how many ATP molecules are formed by complete oxidation?

Given: 42 PBCs

Find: amount of glucose, amount of ATP at complete oxidation.

Solution:

1) In glycolysis, one molecule of glucose is broken down to form 2 molecules of pyruvic acid (PVA), hence, 42 : 2 = 21 molecules of glucose have undergone glycolysis;

2) the complete oxidation of one glucose molecule (oxygen-free and oxygenic steps) produces 38 ATP molecules;

3) the oxidation of 21 molecules produces: 21 * 38 = 798 molecules of ATP.

Answer: 21 molecules of glucose, 798 molecules of ATP

Cell division

(n - number of chromosomes; c - amount of DNA) mitosis meiosis

Interphase	2p2c - 2p4s	6 ■ 10 "^{9} mg 12 10-^{9} mg.	Interphase	2p2c - 2p4c	6 ■ 10 "^{9} mg 12 ■ IO '^{9} mg.
			Prophase i	2p4s	12 10-' mg
Prophase	2p4s	12 10-^{9} mg.	Metaphase i	2p4s	12 10-' mg
			Anaphase i	2p4s	12 10'^{9}mg

Metaphase	2p4s	12 10^{-9} mg.			
			Telophase i	p2c	6 - $10^{'9}$ mg
			Prophase 2	p2c	6 ■ $10^{"9}$ mg
Anaphase	2p4s	12 $10^{'9}$mg			
			Metaphase 2	p2c	6 - $10^{"9}$ mg.
Telophase	2p2c	6 - $10^{"9}$ mg.	Anaphase 2	η 2s	6 - 10^{*9} mg
			Telophase 2	η with	3-10^{-9}mg

The total mass of all DNA molecules in 46 chromosomes of one human tsomatic cell is about 6-10 -9 mg. Determine what is the mass of all DNA molecules in the nucleus during ovogenesis before the beginning of division, at the end of the telophase of meiosis I and meiosis II. Explain the results obtained.

Solution:

1) before division begins during replication, the number of DNA doubles and the mass of DNA is 2 ^{x} 6 ^{x} 10 -9 = 12 ^{x} 10 -9 mg;

2) the first division of meiosis is reduction, the number of chromosomes becomes 2 times less, but each chromosome consists of two DNA molecules (sister chromatids), so in telophase of meiosis I the mass of DNA is 12 ^{x} 10 -9 : 2 = 6 x 10-9 mg;

3) after meiosis II, each nucleus in the cell contains single chromatid chromosomes of haploid set, so in the telophase of meiosis II the mass of DNA is 6 x 10-9 : 2 = 3 x 0-9 mg

Response:

1)before division starts, DNA mass = 12 x 10 -9 mg

2) in telophase of meiosis I, DNA mass = 6 x10-9 mg;

3) In telophase of meiosis II mass = 3 x10-9 mg

Restriction analysis

Allows the identification of biologically important regions on DNA.

Since

A restriction map reflects the location of a particular nucleotide sequence in a given site; comparing such maps for two or more related genes allows us to estimate the homology between them. By analysing restriction maps, it is possible to compare certain sections of DNA from different animal species without determining their nucleotide sequence. It also allows you to see major genetic changes such as deletions or insertions. In this case, there is a decrease or increase in restriction fragments, as well as the disappearance or appearance of restriction sites.

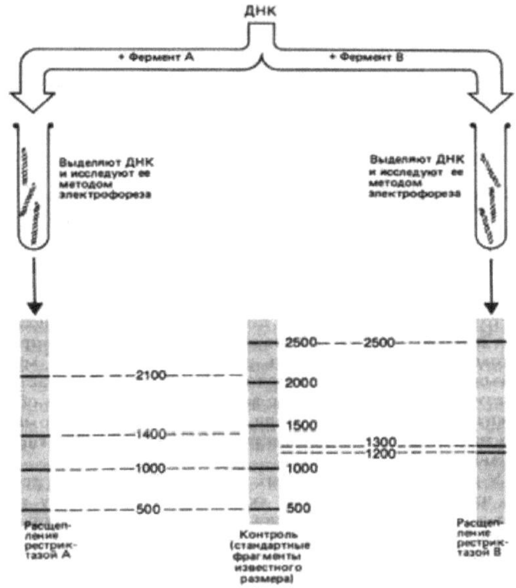

The linear DNA fragment was treated with EcoRI restrictase, BamHI restrictase and their mixture. The reaction products were separated in an agarose gel and stained with ethidium bromide. The results of electrophoresis are shown in the figure. The numbers indicate the approximate fragment sizes in bp. Draw the restriction map of the fragment.

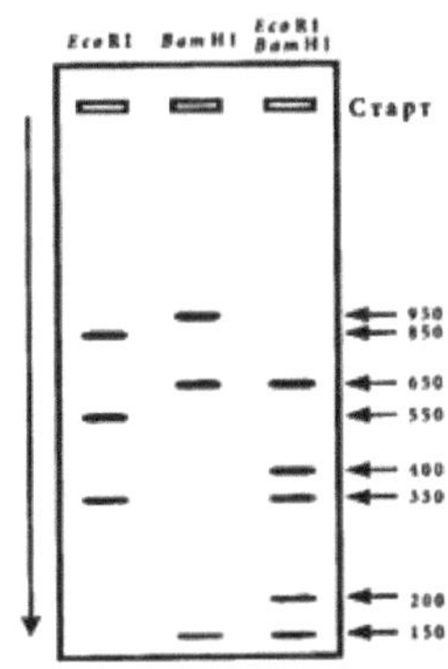

Solution:

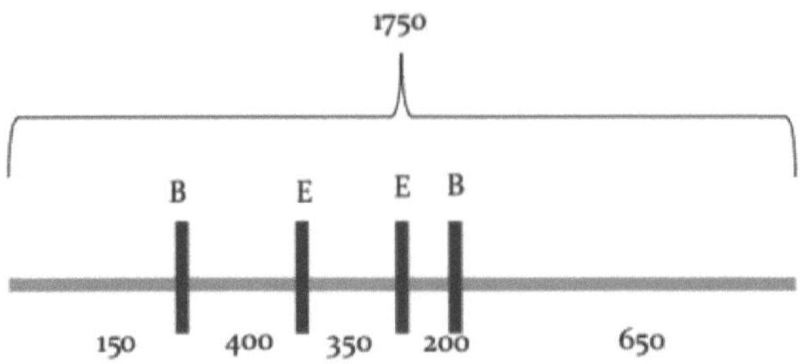

Problems to be solved independently. Solve them after the topic outline and send the solution together with the outline

Tasks for self-solving

1. RNA isolated from tobacco mosaic virus contains 20% cytosine. Can can the percentage of adenine in this RNA be calculated?

2. What is the number of nucleotides contained in the mRNA encoding the polypeptide,
whose molecular mass is 27kDa if the molecular mass of one amino acid in a protein is approximately 100 Da?

3. there is a DNA molecule of the following form:

	1	2	3	4	5	6	7	8	9	10	11
А.	ТАЦ	АТГ	АТЦ	АТТ	ТЦА	ТГА	ААТ	ТТЦ	ТАГ	ЦАГ	ГТА
Б.	АТГ	ТАЦ	ТАГ	ТАА	АГТ	АЦТ	ТТА	ААГ	АТЦ	ГТА	ЦАТ

Where numbers denote the order of triplets, and letters A and B denote individual strands of the DNA molecule. It is known that this DNA provides synthesis of a polypeptide consisting of 5 amino acids. Which DNA strand, from which codon, and in which direction should be transcribed?

1 2 3 4 5 6 7 8 9 10 11

A. TAC ATG ATZ ATT TCA TGA AAT TTC TTC TSAG TSAG GTA

Б. ATG TAC TAC TAC TAA AGT AZT TTA AAG ATG GTA TSAT

4. Complete the table and indicate the correct 5/ and 3/ ends for DNA molecule, mRNA and tRNA.

Ц												Двунитч. ДНК
						Т	Г	А				
	Ц	А				У						мРНК
									Г	Ц	А	Антикод. тРНК
			Триптофан									Аминокисл., включающаяся в белок

5. There is a double-helical DNA molecule that represents a region of a gene:

Suppose transcription starts at nucleotide A in the mRNA, proceeds from left to right, and continues to the end.

Define:

- Sequence of m RNA synthesised;
- Amino acid sequences in a polypeptide;
- how would the structure of a protein change if there was

third CD deletion on a non-coding strand?

5'	Ц	А	Ц	Т	Ц	Т	Г	Ц	Т	Т	Г	Ц	Г	Т	Г	Г	А	Ц	Г	Ц	А	Т	Т	А	А	Ц
3'	Г	Т	Г	А	Г	А	Ц	Г	А	А	Ц	Г	Ц	А	Ц	Ц	Т	Г	Ц	Г	Т	А	А	Т	Т	Г

6. The figure shows the results of electrophoresis of PCR amplified DNA fragments from members of one family (father, mother and 9 children). The father and 6 children (3,5,7,8,10,11) in this family have symptoms of the hereditary disease chorea

Huntington's. The father developed the disease at age 40, and the ages of the children with the first symptoms of the disease are shown in the figure next to the corresponding DNA fragments. What is the probability that children 4,6, and 9 have the disease?

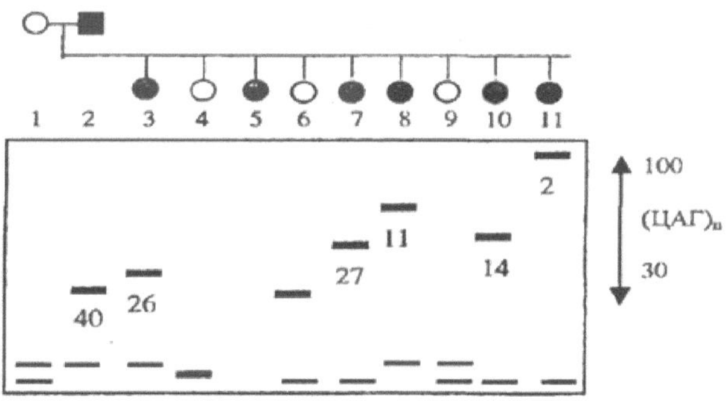

7. The haemoglobin of human blood contains 0.34% iron. Calculate the minimum molecular weight of haemoglobin.
8. Human serum albumin has a molecular mass of 68400. determine the number of amino acid residues in the molecule of this protein.
9. A protein contains 0.5% glycine. What is the minimum molecular weight of this protein, if Mmin= 75.1? How many amino acid residues are there in this protein?
10. During dissimilation, 4 moles of glucose were broken down, of which only 3 moles underwent complete breakdown.

Define:

A) how many moles of lactic acid are formed?

B) how many moles of ATP are synthesised?

B) What amount of energy is stored in them?

D) How many moles of CO2 were formed?
E) How many moles of O2 are consumed?
11. During dissimilation, 13 moles of glucose were broken down, of which only 5 moles were completely broken down.
Define:
A) How many moles of lactic acid were formed?
B) How much ATP was formed in this process?
B) What amount of energy is stored in them?
D) How many moles of CO2 were formed?
E) How many moles of O2 are consumed?
12. Tobacco mosaic virus (RNA-containing virus) synthesises a section of protein with the amino acid sequence: Ala - Tre - Ser - Ala - Tre - Ser - Ser - Ser.
Glu - Met -. Under the action of nitric acid (mutagenic factor), cytosine is converted to uracil as a result of deamination. What structure will the protein section of the tobacco mosaic virus have if all cytidyl nucleotides undergo the above chemical transformation?
Solution (for convenience we use the tabular form of writing down the solution): Using the complementarity principle and the table of genetic code we obtain:

Protein chain amino acids initial)					
andRNASource)					
iRNA (deaminated)					
Protein chain amino acids deaminated)					

To consolidate the theoretical material on methods and techniques of problem solving, there are tasks for independent solution, as well as questions for self-control.
Examples of problem solving
Necessary Explanation:
One step is a full 360o turn of the DNA helix.
One step is made up of 10 pairs of nucleotides
The length of one step is 3.4 nm
The distance between two nucleotides is 0.34 nm
The molecular mass of one nucleotide is 345 g/mol
Molecular weight of one amino acid - 120 g/mol
In a DNA molecule: A+G=T+C (Chargaff's Rule: Σ(A)= Σ(T), Σ(Γ)= Σ(4),

Σ(A+Γ) =Σ(T+^

Nucleotide complementarity: A=T; G=C

DNA strands are held together by hydrogen bonds that are formed between complementary nitrogenous bases: adenine and thymine are joined by two hydrogen bonds, and guanine and cytosine by three.

On average, one protein contains 400 amino acids;

calculating the molecular weight of a protein:

where M min is the minimum molecular mass of the protein,

a - atomic or molecular mass of the component, c - percentage content of the component.

Task. One of the DNA strands has the sequence of nucleotides: AGT ACC HAT HAT HAT HAT HAT CGA TTT HAT HAT ... ... What is the nucleotide sequence of the second DNA strand of the same molecule. For clarity you can use the magnetic "alphabet" of DNA (technique of the author of the article) .

Solution: according to the principle of complementarity, we complete the second chain (A-T,G-C). It looks as follows: TTSA TGG TSTA TGA TGA GZT AAA TGZ.

Task. The sequence of nucleotides at the beginning of the gene that stores information about the protein insulin begins as follows: AAA CAAC DTG DTT GTA GAC. Write the amino acid sequence that begins the insulin chain.

Solution: The task is performed using a genetic code table, in which nucleotides in iRNA (in brackets - in the original DNA) correspond to amino acid residues.

Task. The larger of the two chains of the insulin protein has (the so-called B chain) begins with the following amino acids: phenylalanine-valine-asparagine-glutamic acid-histidine-leucine. Write the sequence of nucleotides at the beginning of the section of the DNA molecule that stores information about this protein.

Solution (for convenience we use the table form of the solution record): since one amino acid can be encoded by several triplets, it is impossible to determine the exact structure of the i-RNA and DNA site, the structure may vary. Using the principle of complementarity and the table of genetic code we obtain one of the variants.

Task. A section of the gene has the following structure, consisting of a sequence of nucleotides: CYG CYC TCA AAA TCA .... Specify the structure of the corresponding section of the protein, the information about which is contained in this gene. What is the effect on the structure of the protein if the fourth nucleotide is removed from the gene?

Solution (for convenience we use the tabular form of writing down the solution): Using the complementarity principle and the table of genetic code we obtain:
When the fourth nucleotide, CD, is deleted from the gene, a noticeable change will occur - the number and composition of amino acids in the protein will decrease:
Task. Tobacco mosaic virus (RNA-containing virus) synthesises a section of protein with the amino acid sequence: Ala - Tre - Ser - Glu - Met-. Under the action of nitric acid (mutagenic factor) cytosine is converted to uracil as a result of deamination. What structure will the protein section of the tobacco mosaic virus have if all cytidyl nucleotides undergo the above chemical transformation?
Task. In Fancomi syndrome (disorder of bone tissue formation) in a patient with urine excrete amino acids that correspond to codons in and -RNA: AUA GUTS AUG AUG UCA UUG GUG AUG AUG. Determine which amino acids are excreted in the urine in Fancomi syndrome, if a healthy person's urine contains the amino acids alanine, serine, glutamic acid, and glycine.
Solution (for convenience we use the tabular form of writing down the solution): Using the complementarity principle and the table of genetic code we obtain:
and -RNA AUA GUTS AUG UCA UUCA UUG GUG UUU AUG
Protein chain amino acids (diseased human) Ise-Val-Met-Met-Ser-Leu-.
Val-Ile
Protein chain amino acids (healthy human) Ala-Ser-Glu-Gly
Thus, in the urine of a sick person only one amino acid (serine) is the same as in a healthy person, the rest are new, and the three characteristic of a healthy person are absent.
Task. The A chain of bovine insulin contains alanine in the 8th link and threonine in the 9th link, serine and glycine, respectively. What can be said about the origin of insulins?
Solution (for convenience of comparison we use the tabular form of the solution record): Let's see what triplets in i-RNA are encoded by amino acids mentioned in the problem condition.

Organism	Bull	Horse
8th link	Ala	Tre
and- RNA	HZU	ACU
9th link	Sulph	Glee
and- RNA	ASU	GSU

Since amino acids are encoded by different triplets, the triplets that differ minimally from each other are taken. In this case, in horse and bull, the amino acids in the 8th and 9th strands have been changed as a result of substitution of the first nucleotides in the triplets and -RNA: guanine has been replaced by adenine (or vice versa). In double-stranded DNA, this would be equivalent to replacing the pair Ts-G with T-A (or vice versa).

Consequently, the differences in the A chains of bovine and equine insulin are due to transitions in the section of the DNA molecule encoding the 8th and 9th links of the A chain of bovine and equine insulins.

Task. Studies have shown that i-RNA contains 34% guanine, 18% uracil, 28% cytosine and 20% adenine. Determine the percentage composition of nitrogenous bases in a section of DNA, which is the matrix for this i-RNA.

Solution (for convenience, we use the table form for recording the solution):

We calculate the percentage ratio of nitrogenous bases based on the principle of complementarity: i-RNA Г У Ц А

34% 18% 28% 20%

DNA (sense strand, readable) G A C T T

28% 18% 34% 20%

DNA (antisense strand) G A C T 34% 20% 28% 18%

The sum of A+T and G+C in the semantic chain will be: A+T=18%+20%=38% ; G+TS=28%+34%=62%. In the antisense chain

(non-coding) chain, the totals will be the same, only the percentage of individual bases will be reversed: A+T=20%+18%=38% ;

D+TS=34%+28%=62%. In both chains the pairs of complementary bases will be equal, i.e. adenine and thymine - 19% each, guanine and cytosine - 31% each.

Task No. 8. On a fragment of one DNA strand, the nucleotides are arranged in the sequence: A-A-G-T-T-T-T-T-A-C-G-T-T-A-T. Determine the percentage of all nucleotides in this DNA fragment and the length of the gene.

Solution:

1) complete the second strand (according to the principle of complementarity)

2) Σ(A +T+T+C+G) = 24, of which Σ(A) = 8 = Σ(T)

24 - 100% => x = 33,4%

8 - x%

24 - 100% => x = 16,6%

4 - x%

Σ(Г) = 4 = Σ(^

3) DNA molecule is double-stranded, so the length of a gene is equal to the

length of a single strand:

12 X 0.34 = 4.08 nm

Task #9. The share of cytidyl nucleotides in a DNA molecule is 18%. Determine the percentage of other nucleotides in this DNA.

Solution:

1 .since Ts = 18%, then G = 18%;

2) A+T account for 100 per cent - (18 per cent +18 per cent) = 64 per cent, i.e. 32 per cent each

Task #10. In the DNA molecule found 880 guanidyl nucleotides, which make up 22% of the total number of nucleotides in this DNA.

Determine: a) how many other nucleotides are in this DNA? b) what is the length of this fragment?

Solution:

1) Σ(Г)= Σ(4)= 880 (this is 22%); Other nucleotides account for 100% - (22%+22%)= 56%, i.e. 28% each; To calculate the number of these nucleotides, we make a proportion:

22% - 880

28% is x, hence x = 1120.

2) to determine the length of DNA, you need to find out how many total nucleotides are in 1 chain:

(880 + 880 + 1120 + 1120) : 2 = 2000

2000 X 0.34 = 680 (nm)

Task No. 11. A DNA molecule with a relative molecular mass of 69 000 is given, 8625 of which are adenyl nucleotides. Find the number of all nucleotides in this DNA. Determine the length of this fragment.

Solution:

1. 69,000 : 345 = 200 (nucleotides in DNA), 8625 : 345 = 25 (adenyl nucleotides in that DNA), £(G+C) = 200 - (25+25)= 150, i.e. there are 75 of each;

2) 200 nucleotides in two chains, so one chain has 100 nucleotides. 100 x 0.34 = 34 (nm)

Task. Which is heavier: the protein or its gene?

Solution: Let x be the number of amino acids in a protein, then the mass of this protein is 120x, the number of nucleotides in the gene encoding this protein is 3x, the mass of this gene is 345 x 3x. 120x < 345 x 3x, so the gene is heavier than the protein.

Task. Haemoglobin of human blood contains 0, 34% of iron. Calculate the minimum molecular mass of haemoglobin.

Solution: Mmin = 56 : 0.34% - 100% = 16471

Task. Human serum albumin has a molecular mass of 68400. Determine the number of amino acid residues in the molecule of this protein.
Solution: 68400 : 120 = 570 (amino acids in a molecule of albumin)
Task. A protein contains 0.5% glycine. What is the minimum molecular mass of this protein if M of glycine = 75.1? How many amino acid residues are there in this protein?
Solution: Mmin = 75.1 : 0.5% - 100% = 15020 ; 15020 : 120 = 125 (amino acids in this protein)

Tasks for independent work

A DNA molecule has split into two strands. One of them has the structure : TAG ACT GGT GGT ACA CDGT GGT GAT GAT TCA .... What will be the structure of the second DNA molecule when this chain is completed to a complete double-stranded molecule ?

The polypeptide chain of one animal protein has the following beginning : lysine-glutamine-threonine-alanine-alanine-alanine-lysine-... With which nucleotide sequence begins the gene corresponding to this protein?

A section of a protein molecule has the following sequence of amino acids: glutamine-phenylalanine-leucine-tyrosine-arginine. Identify one of the possible nucleotide sequences in a DNA molecule.

A section of a protein molecule has the following amino acid sequence: glycine-tyrosine-arginine-alanine-cysteine. Identify one of the possible nucleotide sequences in a DNA molecule.

One chain of ribonuclease (pancreatic enzyme) consists of 16 amino acids: Glu-Gly-Asp-Pro-Tyr-Val-Pro-Val-Pro-Val-Gis-Phen-Phen-Asn-Ala-Ser-Val. Determine the structure of the DNA region encoding this part of the ribonuclease.

A fragment of a DNA gene has the following nucleotide sequence GTC CTA ATC GGA TTT. Determine the sequence of nucleotides and RNA and amino acids in the polypeptide chain of the protein.

A DNA gene fragment has the following nucleotide sequence TCG GTC GTC AAC TTA GCT. Determine the sequence of i-RNA nucleotides and amino acids in the polypeptide chain of the protein.

A fragment of a DNA gene has the following nucleotide sequence TGG ACA GGT TTC GTA. Determine the sequence of nucleotides of i-RNA and amino acids in the polypeptide chain of the protein.

Determine the order of amino acids in a section of a protein molecule if it is known that it is encoded by the following DNA nucleotide sequence: TGA TGC GTT TAT TAT GTSGT GTSG CCC. How would the protein change if the 9th and 13th nucleotides were chemically removed?

The coding strand of DNA has the sequence of nucleotides: TAG TGT TTC TTS TTSG GTA. How will the structure of the protein molecule change if the sixth nucleotide in the DNA chain is doubled. Explain the results.

The coding strand of DNA has the sequence of nucleotides: TAG TTC TTS TTSG AGA. How will the structure of the protein molecule change if the eighth nucleotide in the DNA chain is doubled. Explain the results.

Under the influence of mutagenic factors, the second triplet was replaced by the ATA triplet in a fragment of the gene: CAT TAG TAG GTA TGT TGT

TCG. Explain how the structure of the protein molecule will change.

Under the influence of mutagenic factors in the gene fragment: AGA TAG TAG GTA GTA GTA GGT TGT TcG, the fourth triplet has been replaced by the triplet of ACC. Explain how the structure of the protein molecule will change.

A fragment of an i-RNA molecule has the following nucleotide sequence: GCA UGU AGC AGC AAG CYC. Determine the sequence of amino acids in the protein molecule and its molecular weight.

A fragment of an i-RNA molecule has the following sequence of nucleotides: GAG CCA AAU ATSU UUA. Determine the sequence of amino acids in the protein molecule and its molecular weight.

A DNA gene contains 450 pairs of nucleotides. What is the length, molecular weight, and how many amino acids are encoded in the gene?

How many nucleotides does a DNA gene contain if it encodes 135 amino acids. What is the molecular weight of this gene and its length?

A fragment of one DNA strand has the following structure: GGT ACG ATG ATG TCA AGA. Determine the primary structure of the protein encoded in this chain, the number (%) of different types of nucleotides in the two chains of the fragment, and its length.

What is the molecular mass of a gene and its length if it encodes a protein with a molecular mass of 1500 g/mol?

What is the molecular mass of a gene and its length if it encodes a protein with a molecular mass of 42000 g/mol?

A protein molecule contains 125 amino acids. Determine the number of nucleotides in the i-RNA and DNA genes and the number of t-RNA molecules involved in the synthesis of this protein.

A protein molecule contains 204 amino acids. Determine the number of nucleotides in the i-RNA and DNA genes and the number of t-RNA molecules involved in the synthesis of this protein.

145 molecules of t-RNA took part in the synthesis of a protein molecule. Determine the number of nucleotides in the i-RNA, DNA gene and the number of amino acids in the synthesised protein molecule.

128 t-RNA molecules took part in the synthesis of a protein molecule. Determine the number of nucleotides in the i-RNA, DNA gene and the number of amino acids in the synthesised protein molecule.

A fragment of the i-RNA chain has the following sequence: YYYY UYYY UAU CCC AAC AAC UGU. Determine the sequence of nucleotides on DNA, t-RNA anticodons, and the sequence of amino acids corresponding to the DNA gene fragment.

A fragment of the i-RNA chain has the following sequence: GUUU GAA CYCG UAU CYCU. Determine the sequence of nucleotides on the DNA, the t-RNA anticodons, and the sequence of amino acids corresponding to the DNA gene fragment.

An i-RNA molecule contains 13% adenyl, 27% guanyl, and 39% uracil nucleotides. Determine the ratio of all types of nucleotides in the DNA from which this i-RNA was transcribed.

An i-RNA molecule contains 21% cytidyl, 17% guanyl, and 40% uracil nucleotides. Determine the ratio of all types of nucleotides in the DNA from which this i-RNA was transcribed

An i-RNA molecule contains 21% guanyl nucleotides, how many cytidyl nucleotides are contained in the coding chain of a section of DNA?

If the chain of the DNA molecule from which the genetic information was transcribed contained 11% adenyl nucleotides, how many uracil nucleotides would be contained in its corresponding section of i-RNA?

MIX
Papier aus verantwortungsvollen Quellen
Paper from responsible sources
FSC® C105338

FSC
www.fsc.org

Printed by Books on Demand GmbH, Norderstedt / Germany